Paper Machines

History and Foundations of Information Science
Edited by Michael Buckland, Jonathan Furner, and Markus Krajewski

Human Information Retrieval by Julian Warner

Good Faith Collaboration: The Culture of Wikipedia by Joseph Michael Reagle Jr.

Paper Machines: About Cards & Catalogs, 1548–1929 by Markus Krajewski

Paper Machines

About Cards & Catalogs, 1548–1929

Markus Krajewski

translated by Peter Krapp

The MIT Press
Cambridge, Massachusetts
London, England

© 2011 Massachusetts Institute of Technology

© für die deutsche Ausgabe 2002, Kulturverlag Kadmos Berlin

All rights reserved. No part of this book may be reproduced in any form by any electronic or mechanical means (including photocopying, recording, or information storage and retrieval) without permission in writing from the publisher.

This book was set in Stone Sans and Stone Serif by Toppan Best-set Premedia Limited.

Library of Congress Cataloging-in-Publication Data

Krajewski, Markus, 1972–
[Zettelwirtschaft. English]
Paper machines : about cards & catalogs, 1548–1929 / Markus Krajewski ; translated by Peter Krapp.
   p. cm. — (History and foundations of information science)
Includes bibliographical references and index.
ISBN 978-0-262-01589-9 (hc alk. paper), 978-0-262-55085-7 (pb)
1. Catalog cards—History. 2. Card catalogs—History. 3. Information organization—History. I. Title.
Z693.3.C37K7313 2011
025.3'109—dc22

2010053622

Contents

1 **From Library Guides to the Bureaucratic Era: An Introduction** 1

2 **Temporary Indexing** 9

I **Around 1800** 25

3 **The First Card Index?** 27
Addressing Ideas 27
Data Streams 32
Copy Error: The Josephinian Card Index 34
   Floods 35
   Canals 37
   The Algorithm 38
   Error: Buffer Overflow 42
Paper Flow: Taming, Duration 43
Revolution on Playing Cards 45

4 **Thinking in Boxes** 49
The Scholar's Machine 50
Genealogy: Johann Jacob Moser and Jean Paul 53
Elsewhere 56
Banknotes 58
Balance Sheet 62
   In Praise of the Cross-Reference 63
   On the Gradual Manufacturing of Thoughts in Storage 65

5 **American Arrival** 69
Do Not Disturb—William Croswell 69
Early Fruits and Dissemination 78

II Around 1900   85

6 Institutional Technology Transfer   87
Reformation: Dewey's Three Blessings for America   87
Transfer: Library Bureau   90
   Library Supplies   90
   Standardization   91
   Corporate Genealogy   92
   The Transfer   95
   Product / System / Manufacturing   100
Digression: Foreign Laurels   102
Industry Strategy   104

7 Transatlantic Technology Transfer   107
Supplying Library Supplies   108
   The Library *Ge-stell*   108
   Punch Card   110
The Bridge Enters the Office: World Brain   113

8 Paper Slip Economy   123
System / Organization   125
Universal / Card / Machine   127
   Invalidation   131
   The War of the Cards: Copyrighting the "Card Index"™   133
   Depiction / Decision   135
Summary: Order / Cleanup   139

Afterword to the English Edition   143
Notes   145
References   181
Index   207

1 From Library Guides to the Bureaucratic Era: An Introduction

We wanted to play bureaucratic music.
—Einstürzende Neubauten, *Faust::Mein Brustkorb::Mein Helm*, after Werner Schwab

"Card catalogs can do anything"—this is the slogan Fortschritt GmbH introduces to promote its progressive services in the 1929 volume of the *Zeitschrift für Organisation* (figure 1.1), quite in accordance with its name: *Fortschritt*, progress. The promise offered in the very first phrase of the company's full-page advertisement is a lofty one, and there is more to come:

Card catalogs can *maintain order* among tens of thousands of small and large items in the warehouse management of large industrial plants, they can structure *any number of addresses* in personnel departments, they can *control* the *movement* of hundreds of thousands of people in urban registration offices, they can make themselves useful in the *bookkeeping departments* of commercial offices, i.e. as open account catalogs, *etc. etc. Card catalogs can do anything!* Read more about *Fortschritt*'s innovations in index card equipment in the newly published card catalog *Mobile Notes*.[1]

This book explores the conceptual development of the card catalog from its primal scene to the above "progress." Why *would* this auspicious apparatus be able to do office work such as sorting, addressing, controlling, storing, accounting, and computing? And since the terminology demands situating the card index in a media archeology that examines the universality of paper machines, the questions guiding this study follow the development of (preelectronic) data processing. What makes this promising and supposed jack-of-all-trades a universal machine? As Alan Turing proved only years later, these machines merely need (1) a (theoretically infinite) partitioned paper tape, (2) a writing and reading head, and (3) an exact

Figure 1.1
Fortschritt GmbH: *Karteien können alles!* ("Card indexes can do anything!", Zeitschrift für Organisation und moderne Betriebsführung 3 (23):6 (1929))

procedure for the writing and reading head to move over the paper segments.² This book seeks to map the three basic logical components of every computer onto the card catalog as a "paper machine," analyzing its data processing and interfaces that may justify the claim, "Card catalogs can do anything!"

This statement marks the preliminary closing point of a development that culminates in the bureaucratic apparatus called "card catalog" (from around 1920 into the 1960s).³ Numerous new magazines for office supplies and office organization document its wide circle of influence. This book seeks to trace how the card catalog manages to establish itself on every desk around 1930, and how it manages to move to the center of organizational attention. Although the card catalog may appear rather insignificant next to the delicately imposing typewriter, it stubbornly claims its place by its promise of universality. What achievements is this claim based on? Against what background does the paper machine stand out? What roots does this system of recording stem from?

The assertion of a universal paper machine (and a first search for its origin and development) raises the suspicion that this apparatus has its model and predecessor in boxes of paper slips as used by libraries. The basic assumption is that the genealogy of the card catalog as a storage technology includes several technology transfers between discourses: that of the library and that of efficient management. The ubiquitous presence of the card catalog on desks between World War I and World War II owes to a shift of this concept from library to office.

When dealing with such a transfer, it would be insufficient to mark only beginning and end, sender and receiver. Rather, what is crucial is the way this transfer occurs, including any disturbances, changes, stoppages, irritations, and detours. The method can be described as one that satisfies the basic operations of a *universal discrete machine*: storing, processing, and transferring data.⁴ What differs here from other data storage (as in the medium of the codex book) is a simple and obvious principle: information is available on separate, uniform, and *mobile* carriers and can be further arranged and processed according to strict systems of order.

This technology transfer harks back to a primal scene, even though at first it is limited to libraries and closely linked learned discourse. Polymath Konrad Gessner stands at the beginning of this history of the card catalog. He is probably not the inventor of the technique of cutting up pieces of information on paper so as to (re)arrange them more readily. However, his explicit description of this process in 1548 may constitute the earliest account of conveniently generating extensive lists in alphabetical order. Therefore, chapter 2 will be devoted to a detailed description of Gessner's paper machine, pursuing the tradition of this exemplary, highly recommended device in early modern times and the Baroque era, especially in the art of excerption.[5]

Starting off part I, "Around 1800," chapter 3 traces a discontinuity, a break in the application of Gessner's procedure. When the technique goes from provisional to permanent, an unintended and yet consequential turn takes place, giving rise to the first card catalog in library history in Vienna around 1780. For the card catalog to become the librarian's answer to the threat of information overload, precise written instructions that can integrate untrained staff into the division of labor are decisive. Here, I also briefly digress and examine two coinciding addressing logics: In the same decade and in the same town, the origin of the card index cooccurs with the invention of the house number. This establishes the possibility of abstract representation of (and controlled access to) both texts and inhabitants.

Chapter 4 stays with the turn of the century around 1800, drawing a necessary distinction between the many-to-many technology of the library and the idiosyncratic order of learned excerpt collections. In other words, this differentiation emphasizes a server concept on the one hand and a workstation philosophy on the other. The first card catalog is not only the product of a collective process, but also suited for (and conceived by) many. In contrast to this multiuser system, the idiosyncratic machine taking shape in a scholar's box of excerpts withdraws from outside inspection. No mediating role is needed; its internal system is allowed to be incomprehensible to outsiders. This is a form of data protection we can trace from the peculiar scholarly excerption techniques of Johann Jacob Moser to those of Jean Paul.

Chapter 5 is devoted to the transfer of old European library technology to the New World. On the one hand, the box of paper slips reaches the East Coast of the United States through librarians who study in Europe and then apply the practice to the cataloging of their growing collections in the course of the nineteenth century. On the other hand, the United States has its own home-grown technique. In 1817, William Croswell's unfortunate project of devising a comprehensive catalog for the Harvard College Library marks the birth of the American card index—out of a spirit of sloth.

Part II, "Around 1900," focuses on the discursive transfer between library and office, which does not emerge from European libraries. While the latter remain mired in the quarrel over cataloging versus classified shelf arrangement, the initiative is taken by the American Library Association (particularly by Melvil Dewey as a protagonist), kicking off a powerful technology transfer, described in chapter 6, between the institutions of knowledge management and those of business. Dewey's *Library Bureau* not only carries the aim of institutional transfer in its name, but soon develops from a one-man business to a significant corporation.

Chapter 7 is devoted to the effects of this storage technology in Europe, as institutions take up the knowledge of the application and its assumptions (above all, that of universal standardization) to launch card catalogs as an indispensable basis of their own work.

Finally, in chapter 8, list-sorted management on an index card basis is coupled with the organizational discourse of scientific management, which discovers the card index as an economic optimization tool and develops it into an instrument of rationalization. Around 1920, advocates of modern office organization expect great gains from the card index for batch processing, and thus a simple former library technology turns into a new paradigm for book-keeping. The end of this history is marked by a media change that propagates the card index as a new vehicle of civilization that will render the book obsolete.

The present study, organized implicitly according to Shannon's model of communication in channels and streams, also follows that same model in the third transfer—namely, the source of distortion between sender and receiver—only to conclude that noise will carry the day: hence, I would like to include moments of disruption and sources of distortion that endanger the transfer and introduce the element of risk into this use of cards.[6] What is the systemic position of this noise? What safeguards can succeed in limiting the informational entropy of the card index, its irreversible chaos? Which sensitive spots are irritated again and again so as to interrupt and end transfers? For it seems inevitable that the history of the card catalog should come to be read as a history of multiple failure.

The discursive transfers between institutions and also within the card catalog configure the history of the card catalog. In this book, I seek to write this history *from the material*, thus allowing many voices to be heard, naturally at the risk of discordant polyphony. However, as the task consists of tying together episodes involving an arrangement of paper slips and their respective links, I will allow index cards to lead the way.[7]

This very box of index cards may leave some issues aside owing to the limited scope of the study—and sometimes simply owing to lack of information. Its first deficiency is thus an inability to write a universal story of index cards. Therefore, the trajectory does not begin at the dawn of history, and does not describe Mesopotamian, Egyptian, Greek, or Roman methods of cataloging stored texts. Also excluded is the famous library of Alexandria with its equally famous librarian Callimachus, who affixed inventories of texts on clay *pinakes* on the shelves, as well as the Roman *laterculi* or administrative registers.[8] For neither are paper machines—both use different materialities, the by far more valuable and costly papyrus on the one hand and clay on the other.[9] Instead, we begin card index history in the sixteenth century, with an alleged origin that, all misgivings about choosing one such entry notwithstanding, can serve as a provisional starting point for paper catalogs, even though one could have started earlier.

Nor is this study able to remedy a lack that Foucault proclaimed in a footnote: "Appearance of the index card and development of the human sciences: another invention little celebrated by historians."[10] Although one development, the "make-up of the human sciences," serves as a methodological example for this book, a direct connection to the appearance of the index card could not be made unambiguously. The plan had been to fulfill the promise of that footnote and develop the transfer between librarians and businesspeople around 1890 in an appendix to the evolution of the index card. Yet this plan fell victim to lack of space. In such a chapter one would have found not only the development of the materiality of the index card from paper scrap to cardboard. (This episode now runs implicitly through the description of progressive standardization.) Foucault's remark also makes a connection around 1833, citing Bonneville, between card catalog and mercantile directory. Unfortunately, a monograph on the technical media of commerce has not yet been written, or at least it is as yet unknown to the card index at work here. That is why the circulation of money and ideas is only outlined so as to describe an isomorphic logic of the representation of slips of paper and banknotes, pointing to their merging in card/data/banks.

One more thing ought to be explained in advance: why the card index is indeed a paper machine. As we will see, card indexes not only possess all the basic logical elements of the universal discrete machine—they also fit a strict understanding of *theoretical kinematics*. The possibility of rearranging its elements makes the card index a machine: if changing the position of a slip of paper and subsequently introducing it in another place means shifting other index cards, this process can be described as a chained mechanism. This "starts moving when force is exerted on one of its movable parts, thus changing its position. *What follows is mechanical work taking place under particular conditions. This is what we call a machine.*"[11] The force taking effect is the user's hand. A book lacks this property of free motion, and owing to its rigid form it is not a paper machine.

One more word about some of the guiding metaphors and analogies in this study, as they may seem strange, at first glance, to an audience beyond certain academic circles familiar with the premises of New German Media Theory.[12] Every metaphor or allegory, any analogy or comparison has its limits. Catachresis in rhetoric is a failed transfer, a juxtaposition of incongruous elements. The present study consciously takes this risk when using the term "book flood"—though it is an old phrase. When index cards are compared to bank notes or business cards, it is not to claim equivalence or similarity in their function. Rather, the goal is to point to structural similarity without denying differences. The risk of an imperfect figure of speech is taken because metaphors, allegories, analogies, and parallels harness a specific power of insight this study intends to deploy to good effect. For on the one hand, as Donna Haraway and others assert, language—even a highly technical one as in the case of computer programming or mathematics—is figurative.[13] Metaphors are unavoidable in any description, so they should be used in full consciousness of their effect. On the other hand, their use is based on the assumption that appropriate metaphors open our eyes to more than denotation: they produce a surplus of meaning that stimulates thought. This epistemological payoff is contained in the description—in short, the necessary catachresis of historiography activates insights that are called forth by the various connotations of the metaphors chosen.

Thus, readers ought not be too surprised by references that may appear peculiar at first. They will find correspondences between index cards and bank notes, house numbers and book shelving, card catalogs and Turing machines, masses of books and their description as waves of a flood. These images are called upon consciously, at times by way of historical quotations. Even if it is clear that a card catalog does not perfectly resemble the digital calculator or computer, I maintain that the card catalog is *one* precursor of computing.[14] On the software side, the components of the catalog and its function correspond to the theoretical concept of a *universal discrete machine* as developed by Turing in 1936, with a writing/reading head (or scriptor), an infinite paper band partitioned into discrete steps (or slips), and an unambiguous set of instructions for reading and writing data.[15] Moreover, on the hardware side there is a line of industrial development from library technology directly to the producers of early computing installations, pointing to the technology transfer from the catalog card to the punch card and on to modern storage media. To protect the analogies suggested here from the threat of mere speculation, and to restrict them carefully to whatever valuable insight they can provide, such passages are annotated so as to provide readers with information about the limits and extensions of the comparisons made.[16]

The arc of this history sets out with a library guide, not in the sense of an agent that shows the way around the library, but in terms of marking the place where cataloging principles mature in the form of the card index, leading to other applications. And it ends in the age of the office, an era of productivity minus the concept "service," and of office devices minus electricity.

2 Temporary Indexing

With the invention and spread of printing with movable type, a complaint arises in the learned reading world. It is the *book flood*, always a nautical or irrigation metaphor, that has a disturbing effect on readers in the newly established privacy of their studies.[1] "There are so many books that we lack the time even to read the titles," notes the Italian bibliographer Anton Francesco Doni in 1550, already pointing toward the *increasing reading* of titles and footnotes as a principal reaction to too many texts.[2] The explosion of written material after the introduction of the printing press brings a lot of attention to the library, which it did not garner in medieval times. "The Middle Ages read and wrote little," a 1920s library history summarizes laconically.[3] In the medieval canon, *the one book* dominates the selection and reading of all other texts, so that the medieval library yields to the domination of biblical order and selection patterns; inventories comprise between a few dozen and a couple of hundred volumes, and it is merely for the sake of inventory control that they are listed.[4] Only when the library is inundated is the need to deal with all this material recognized. In view of this difficulty, it is hardly surprising that efforts are made, as a kind of emergency response, not only to sort through the masses of books, but also to furnish an exact and thematically purposeful orientation. Claiming to navigate through the book flood, the Swiss doctor, polymath, and mountaineer Konrad Gessner[5] (1516–1565) publishes the *Bibliotheca Universalis*; the first two of three planned volumes come out in Zurich in 1545 and 1548—taken up here only in terms of library innovation.

The first volume of the *Bibliotheca Universalis*, which Gessner starts to sketch at the age of 25, consists of a bibliography of around 3,000 authors in alphabetical order, describing books in terms of content and form, and offering textual excerpts.[6] The bibliography lists over 10,000 texts, breaking with the legacy of listing books in a specific way. In contrast to previous catalogs, the *Bibliotheca Universalis* undertakes an appraisal of the content of the holdings. Conventional catalog lists mostly tended to originate within the scope of annual inventory checks. Gessner, however, examines every single book meticulously to gather complete specifications of format, title, authors (provided they are named or discoverable), place of publication, and year of publication.[7] Then he appends a content description. Hence, Konrad Gessner can rightly count as the father of the modern bibliography.[8] Earlier catalogs, mostly limited to theological subjects, did not make a distinction between source and representation, between their description and quotations from other catalogs, and hardly ever sorted their lists alphabetically.

The second volume of the *Bibliotheca Universalis*, published in 1548 under the title *Pandectarum sive Partitionum Universalium*, contains a list of keywords, ordered not by authors' names, but thematically. This introduces a classification of knowledge on the one hand, and on the other hand offers orientation for the novice about patterns and keywords (so-called *loci communes*) that help organize knowledge to be acquired.[9] "Many scholars accumulate such *loci* in the course of their varied reading in *Commentarios* or *chartaceos libros*, sorted according to titles and classes, either in alphabetical order, as Dominicus Nanus did in the *Polyanthea*, or according to genres or divisions of philosophy or other principles."[10] The order of the commonplaces at first follows a tree structure, carefully partitioned into 21 main classes; remarkably, grammar appears first and theology last (see figure 2.1).[11] Each of these 21 classes and chapters is articulated into paragraphs and separated by the symbol "¶." Gessner writes a preface to some, with rather extensive comments.[12] A list of authors' names follows, along with titles referring to the first, alphabetical compilation, directing readers to the description of texts in the first volume on corresponding subjects.[13]

Only the twentieth chapter on medicine (which Gessner practices to make a living) is not published, presumably because the proceeds of the first part of the *Bibliotheca Universalis* failed to meet the expectations of the publisher, Christoph Froschauer. Others suppose Gessner may have been so dissatisfied with the results that he refrained from publishing them.[14] In 1549, one year after the first nineteen *Pandectae*, the twenty-first is finally printed, in God's name (= theology). Theological commonplaces

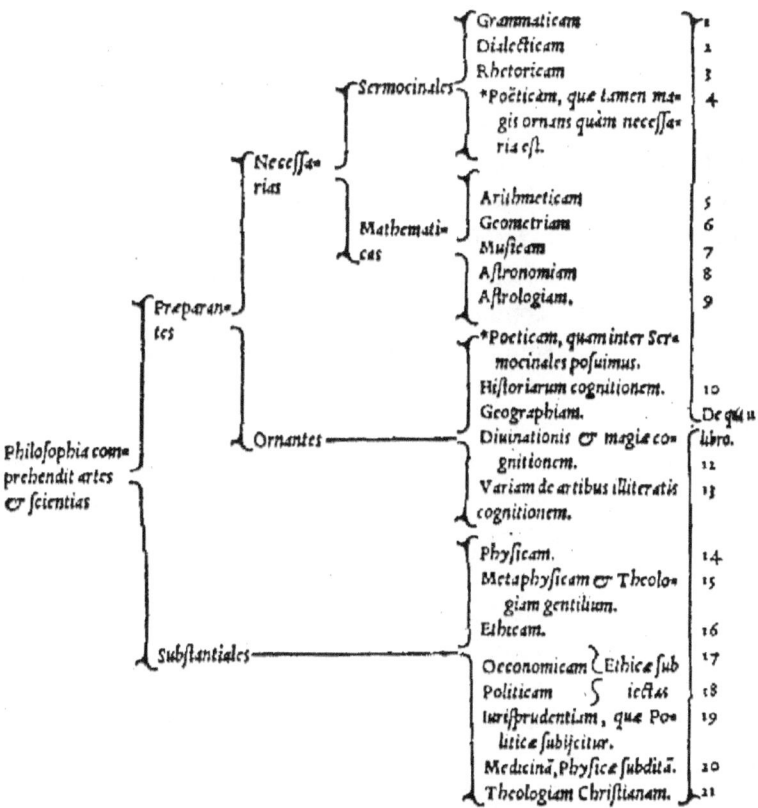

Figure 2.1
The classification system of the *Pandectae* in Gessner 1549, following a tree structure. (From Gessner 1549, page after the cover)

Alcorani cribratio	123 a c
Allegoriæ scriptæ	8 d c
Allegoriæ poeticæ	61 b
Alexander magnus	140 c
Alexandria	150 b
Algorithmus	75 b
Alimenta	344 b
Almanach	94 b
Almageſtum	88 a

Figure 2.2
Selection from the index of commonplaces. (From Gessner 1549, fol. aa^v.)

are followed by an apologetic correction of the editorial outline, and an edited summary of the planned third part, which remained unrealized for lack of money.[15] It was supposed to have been an extensive, alphabetically ordered index to the systematic *locorum communes* of the entire twenty-one chapters of the *Pandectae*. Instead, the appendix lists approximately 4,000 general entries on a mere 26 folio pages, comprising only a small share of potential *locos*, compared with 4,300 juridical commonplaces for *Iurisprudentia*.[16] The slim index couples every commonplace in strict alphabetical order with a reference to the respective page in the *Pandectae* (see figure 2.2).

With this aggregation of commonplaces, the *Pandectae* sought to provide references to and overviews of the alphabetically ordered first part of the *Bibliotheca Universalis*. The assignment of commonplaces to corresponding fields arises from the way that the notes are made; that is, titles are not fixed in a predetermined pattern. Rather, topics are complemented and enhanced according to individual need.[17]

Perhaps the earliest description of such a practice is Gessner's compilation of alphabetical and systematic lists by hand over many years of solitary nocturnal work.[18] The basis of his activity is a careful and comprehensive collection of literature, and Gessner does not shrink from recording the remote and esoteric. "I have gathered material from everywhere: from catalogs of printers, of which I have numerous from different areas; from library directories as well as from libraries themselves, public as well as private ones in Germany and Italy that I carefully researched myself, from letters of friends, from reports of scholars and, finally, from author's catalogs."[19] What does one do with the material thus accumulated? Gessner willingly offers guidance in the corresponding systematic entry of his *Pandectae* called "De Indicibus Librorum."[20] First of all, one should read books that offer an index and register from which to compile one's own indexes. Chapter headings are recorded for use as *locos communes*. This procedure suggests that one may understand the *Bibliotheca Universalis* itself as an index of indexes. The processing of excerpts follows the simplest algorithm:

1. When reading, everything of importance and whatever appears useful should be copied onto a good sheet of paper.

2. A new line should be used for every idea.

3. "Finally, cut out everything you have copied with a pair of scissors; arrange the slips as you desire, first into larger clusters which can then be subdivided again as often as necessary."[21]

4. As soon as the desired order is produced, arranged, and sorted on tables or in small boxes, it should be fixed or copied directly.[22]

Fixing the mobile paper scraps on a sheet means fastening them with glue. This method of attachment, though, should always allow rearrangement, either by use of a water-soluble glue or by some system allowing easy subsequent insertions and shifts. This method requires a special type of book with a guiding thread. As on a weaving loom, the threads stretch over the page, so that two rows of paper slips can be inserted and supported by paper rails (figure 2.3). This device can support up to 100 pages before it becomes unstable.[23]

This procedure describes a hybrid card catalog in book form. And it is by no means pure theory; in fact, Gessner emphasizes its practicability with reference to Georg Joachim Rheticus, "the most excellent mathematician of our time," recommending this model to the entire scholarly world.[24]

Whether they need to write or to give lectures, they may arrange the accumulated raw material for their paper in this way: Either they have recently collected material, or they arrange material accumulated on slips of paper according to thematic aspects for reuse, so they can take out paper slips for the treatment of the respective object, selecting from the many cards those that are best suited for the present subject. Using small needles, they fixate the slips in the desired order for the respective lecture and write down what seems appropriate, or use it according to desire; finally, they restore the slips of paper to their place for reuse.[25]

The recommended care and safekeeping of paper slips in their box builds on reuse, unlike the procedure for indexing, which is a temporary one.

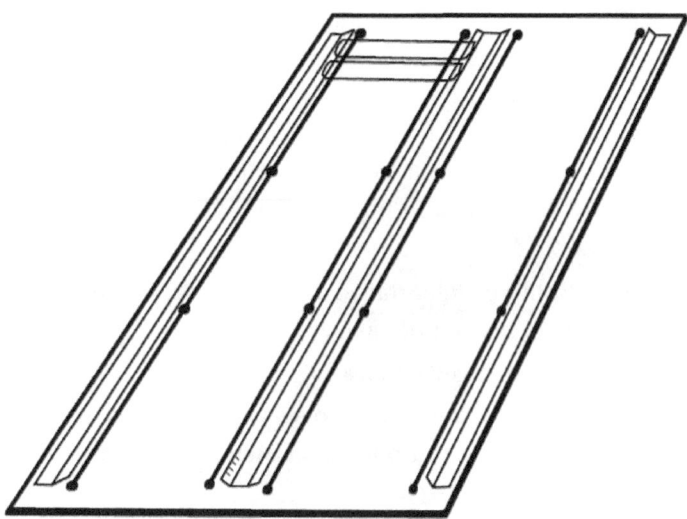

Figure 2.3
The fixation of paper slips. (From Wellisch 1981, p. 12.)

Posthumous praise of the suggested approach is unequivocal: "I can certainly testify this much: that numerous learned men are known to me who like to apply this in their studies."[26] Paper slips can be rearranged again and again in different orders, serving as a basis for text production. This method allows the writing of several books at the same time. And finally, it is even possible to cut books up to save oneself the trouble of copying.[27]

Behind this order of paper slips that guarantees mobility and rearrangement, one can recognize the same economy of signs that a century earlier contributes to a major paradigmatic shift. Johannes Gutenberg's invention of the printing press not only forges most obviously associations of typesetting, steel models, pouring mechanisms for individual letter types, special alloys, and composing sticks for setting lines of type.[28] As a marginal yet indispensable aid, a new tool for filing and storing of the individual pieces of type is introduced: the *type case*. "The technical core of Gutenberg's invention consists in dissolving the articulated sequence of words and letters into their components so as to deploy them as isolated single elements over and over again."[29] In contrast to printing with full-page wood carvings (so-called *block books*), typography owes its potential for recombination to individual precision-cast letters of a special alloy of lead and a little antimony. The typeface turns into a technical element, and its random combination and rearrangement turns handwriting into standardized script. This paves the way for a new economy of textual production that overcomes the laborious and error-prone manual copying and lowers production expenses. Henceforth, bookmakers can count on reuse.

The types for book and newspaper layout lie in wooden cases with about 110 fields for German and 160 for Roman letters, i.e. Latin, English, French &c.; the larger number is required by accents and capital letters. The size of the fields is adapted to the more or less frequent occurrence of each letter, and their position is determined by easy handling. The type case is supported by a chest-high shelf with drawers. The typesetter stands in front of the shelf, holding a metal composing stick in his left hand which forms a level small box, open on two sides, into which the typesetter with his right hand fits the type from the compartments of the box, thus composing them in lines.[30]

The most frequently used type elements—"e," "a," spacers—are in the most quickly accessible places (figure 2.4). The arrangement of the box thus already obeys what would much later be called the Taylorist principle of the shortest motion sequence. After printing, each letter is returned to its drawer, awaiting further use in the production of new pages. Thus, the type case with its compartments for discrete elements is the driving force

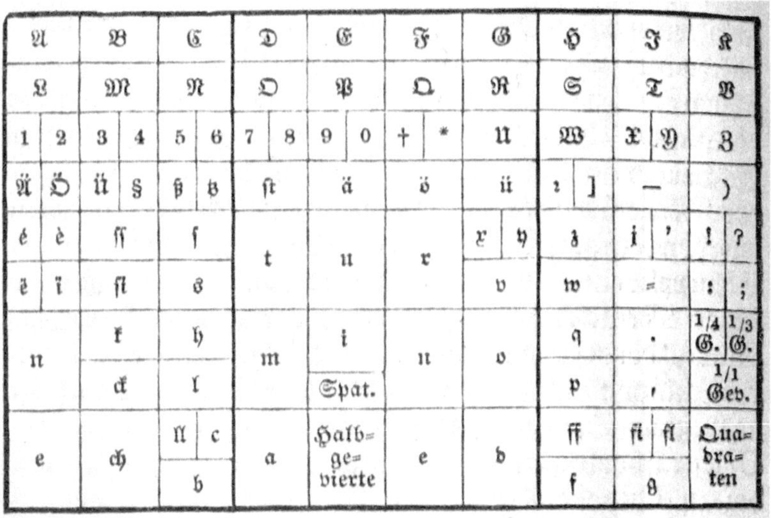

Figure 2.4
Character distribution in the type case. (From Meyers Großes Konversations=Lexicon 1906, vol. 3, p. 528.)

behind segmented storage as suggested by Gessner for bibliographical units on slips of paper in subject or alphabetical order, for the generation of new texts through recombination. Whether for the production of catalogs or as building blocks for learned excerpts, the carefully stored slips of paper allow long-lasting use. For both type case and card catalog, it is essential to keep the respective materials in a flexible form so as to enable the creation of ever new and different arrangements.

This "method of generating indexes in the shortest time and in the best order"[31] is the earliest explicit description of how to store what one has read and found worth keeping, arranging it in different ways, and keeping it thematically retrievable.

It is an important fact that the *Bibliotheca Universalis* addresses a dual audience with this technology of indexing: on the one hand, it aims at librarians with its extensive and far-reaching bibliography; on the other hand, it goes to didactic lengths to instruct young scholars in the proper organization of their studies, that is, keeping excerpted material in useful order. In this dual aim, the *Bibliotheca Universalis* unites a scholar's communication with library technology, before these directions eventually branch out into the activity of library professionals on the one hand and the private and discreet practices of scholarship on the other hand.

However, learning and library were to remain closely connected for a few centuries. Librarians tended to be established scholars and vice versa. Two paths will be briefly illuminated in this context. One will lead to Gottfried Wilhelm Leibniz. The other one has fallen into oblivion: Hugo Blotius (1534–1608), a Dutch Calvinist, lawyer, rhetorician, later a baron and converted Catholic—and the first librarian of the Royal Court library at Vienna.[32] What Leibniz and Blotius have in common is that they tackle adaptations of the excerpting methods suggested by Gessner's influential *Bibliotheca Universalis* to provide efficient catalogs for the book collections entrusted to them in Wolfenbüttel and Vienna, respectively.

Following the chronology of the history of the index card, we first turn to Hugo Blotius, whose term in office in the dusty halls of the Vienna court library lasted from 1575 to 1608. "Dear God, what a state we found them in [...] last July! How neglected and disorderly everything appeared—so much mold and decay everywhere—so much damage caused by moths and bookworms—and everything covered in cobwebs!...When the windows that had been shut for months, never permitting even a ray of sunlight onto these sad and slowly decaying books, were finally opened—indeed a flood of stale air poured out!"[33] Blotius's first approach to this site begins with a catalog made by his own hand, taking stock of approximately 7,700 volumes.[34] When he compiles a systematic overview of this partial inventory, it is in order to demonstrate his ideas of proper cataloging based on what Gessner suggested and practiced in the *Pandectae*, as well as to summarize the urgently needed intelligence on the enemy in the comprehensive Turcica catalog (1603).[35] Thus, he explicitly treats his catalog project as an example to show how future catalogs could be handled. Nevertheless, nobody follows in Blotius's footsteps. Catalogs on different countries and on disparate topics remain in the planning stage, organized in terms of neither scientific nor philosophical order, but instead expressing an interdisciplinary tendency, with special catalogs on joy, grief, and so on.[36]

In 1657, the first practitioner of nonhierarchical indexing, Joachim Jungius (born 1585), dies in Hamburg after compiling approximately 150,000 slips of papers with accumulated knowledge, bound and sorted according to the most minute details and building blocks and without registers or indexes, let alone reference systems.[37] After his death, his pupils and executors discuss adequate scientific excerpting of texts. The debate culminates in *De Arte Excerpendi: Of Scholarly Book Organization* by Vincentius Placcius. It offers an overview of contemporary procedures, instructions on regular excerpting, and an extensive history of the subject. Placcius expressly warns against a loose form of indexing as pursued by Jungius.[38]

Instead of jotting down every note, thought, observation, excerpt, and reading on a single slip of paper and leaving them there, Placcius recommends grouping them, by dint of appropriate *libros excerptorum* (see figure 2.5),[39] excerpt books that follow the pattern that had guided the design of registers and bound library catalogs since Konrad Gessner; Placcius cites the entire section "De Indicibus Librorum" from the *Pandectae*.[40] Perhaps his strong vote for the bound form of this storage technology and against Jungius's method is a way of warning against lifelong knowledge accumulations that merely gather treasures without being recombined and published as new books. Jungius, keen on including new resources, delays his own publications time and again, leaving them unfinished or simply as raw paper slip potential in storage, on call.[41]

Vincentius Placcius's writing provides further insight regarding the appearance of boxes of excerpts that will prove relevant to the history of

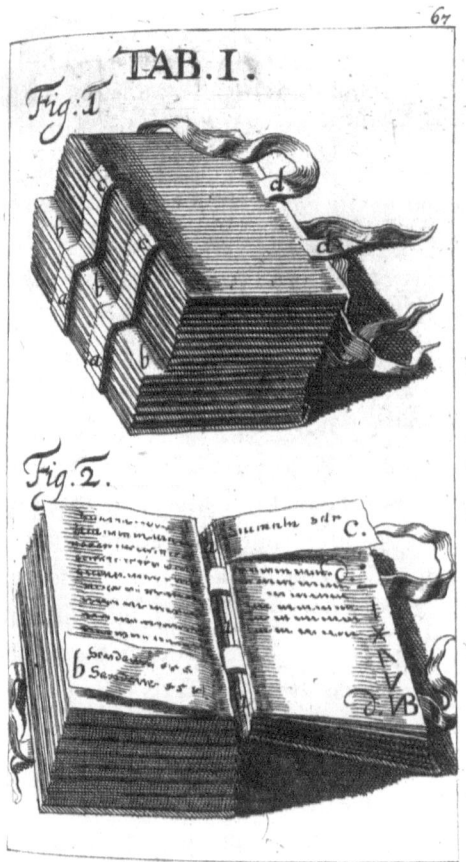

Figure 2.5
Hybrid card index in book form. (From Placcius 1689, p. 67.)

card index development. In the appendix to his work on the art of excerpting, Placcius incorporates word for word an anonymous text from 1637 that introduces two different card index box constructions, including illustrations (see figure 2.6).[42] With their aid, one no longer need worry about errant slips of paper or the difficulty of lack of excerpt mobility. Both variations show internal, replaceable wooden strips that can fix paper slips, either hung with "needles" or "pierced on the corners and thus stitched" (see figure 2.7).[43]

Regarding the legacy of this piece of office furniture, it should be noted that it serves as a model for the construction of an excerpt cabinet

Temporary Indexing

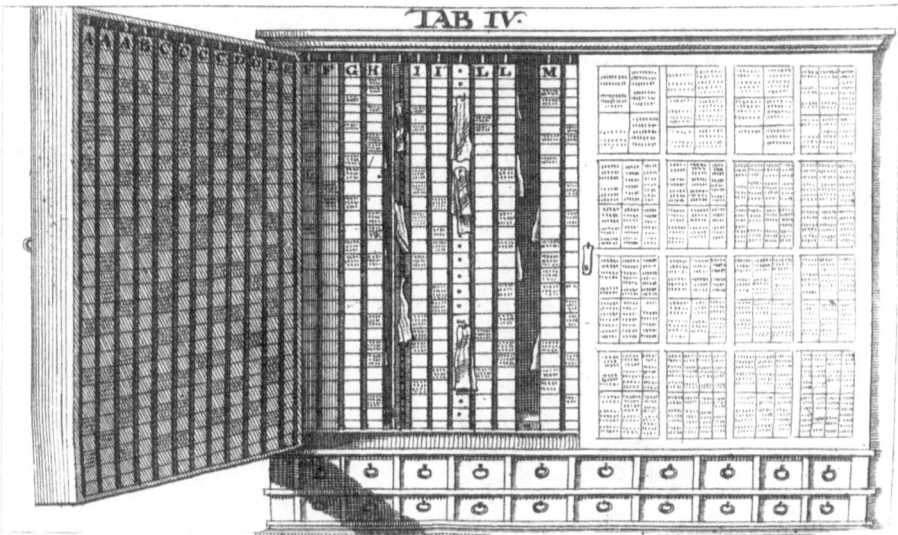

Figure 2.6
Excerpt cabinet. (From Placcius 1689, p. 152.)

constructed for a secretary named Clacius in Hannover, which is acquired after his death by none other than Gottfried Wilhelm Leibniz. From 1676 onward, he follows an excerpting practice that directly refers to Jungius (via one of his students).

Regarding Leibniz's Excerpt Cabinet

He wrote on slips of paper whatever occurred to him—in part when perusing books, in part during meditation or travel or out on walks—yet he did not let the paper slips (particularly the excerpts) cover each other in a mess; it was his habit to sort through them every now and then.

We know, not just since Nietzsche, that good ideas are born while walking. Thus, Leibniz always carried paper with him, for he knew that the administration of fleeting thoughts makes the paper slip indispensable. He transferred his ideas, excerpts, and notes onto paper slips for later sorting and revisiting:

He bought a special cabinet to store his excerpts; [...] Leibniz was in the habit of writing his excerpts on special sheets or slips of paper, and it is likely that he adopted this method from Martino Fogelio (who edited the Jungiana) [...] Yet his method stems from Joach. Jungius. Thus, he also maintained his library according to topical order, without regard for different formats. Leibniz imitated this (presumably old

Figure 2.7
The hook in the excerpt cabinet. (From Placcius 1689, p. 155.)

libraries were also arranged this way) and applied it to his own private library. He had small labels stuck to repositories that indicated what was contained on every shelf.[44]

Our lively survey of the early history of the scholar's box of paper slips may be summarized as classification systems using both software, meaning the question of what principles can order scientific and library data, and hardware, meaning long-term storage devices: (1) the book (Gessner); (2) the nearly immobile, heavy piece of furniture, as yet unnamed, but, as figure 2.6 clearly shows, a kind of *card index cabinet* (Placcius); and (3) the loosely sorted pile of papers on a table, at times filed in envelopes (Jungius).

Up to this point, we still lack a proven set-up that later houses index cards: Gessner only briefly mentions an index card filing box. It joins the scrapbook as a rival product only about 100 years after the *Bibliotheca Universalis*.[45] In 1653, Georg Philipp Harsdörffer completes the idea of an excerpt collection as an ordering of paper slips in a box with twenty-four drawers in alphabetical order. "The content is indexed according to A b c, and for this purpose it is very useful to have one box with twenty-four drawers, each of which bears a letter. When one wants to index, one writes the content on appropriate pieces of papyr and puts them in their letter drawer: later, one takes them out, sorts one letter after another, and either glues the little papyri together, or copies them again."[46] Nevertheless, in contrast to Jungius's use of loose slips, this procedure is not intended as permanent storage; it merely seeks to provide temporary relief following the methods of the *Bibliotheca Universalis* on compiling indexes.[47]

If Leibniz's application in the year 1680 for employment at the Vienna court library had not failed for religious reasons,[48] he might have prevented—if not entirely remedied—the increasing fragmentation of the library catalogs since the time of Hugo Blotius. Alas, the court library remained in a terribly neglected state: books were "piled on top of each other and some of them half decayed, others by *ipsum non usum* completely disintegrated."[49] Nonetheless, Leibniz proves circumspect when first hired in Hannover, and later as librarian at the famous Baroque library of Wolfenbüttel, in the completion of the cataloging projects he is entrusted with.[50] Between 1691 and 1699, again following Gessner's procedures, he lists the extensive holdings of the Wolfenbüttel library with the aid of two secretaries writing, an assistant cutting, sorting, and gluing, and two theology students copying: a directory that was to remain, into the twentieth century, the only general author catalog of the Duke August Library.[51]

"The library is the treasury of all wealth of the human mind in which one takes refuge," Leibniz writes in a letter to Friedrich of Steinberg in October 1696.[52] Although around 1700 it becomes increasingly difficult to gain access to this wealth—the inventories of the mostly aristocratic libraries have begun to grow quickly, thanks to the Baroque passion for collecting—in 1699 Leibniz can await the book flood as calmly as any reader's request.[53] "Not every visitor asks: Do you have Montaigne, or Rösel? From time to time someone says: What liturgical or economic works do you have? This necessitates a logical or scientific catalog."[54] In Wolfenbüttel, one receives a brief and reliable answer to the question of how one might find something in the massive inventory: with the aid of an ensemble of alphabetical catalogs and *indices materiarum* that obey as their highest principle "that one finds the popular books easily."[55] Nevertheless, at the beginning of the eighteenth century it is by no means self-evident that a library should own a directory of its holdings. To find a book on a certain subject, one usually follows the classified shelving of books. At the outset of his library activity in Wolfenbüttel, Leibniz sketches a detailed plan, aiming to tackle the pitiful mess this famous collection is in. For a library without a catalog, as Leibniz put it in his *Consilium*, resembles the warehouse of a businessman who cannot keep stock.[56] If the purpose of a businessman is garnering profits from his products, deploying certain technologies such as double-entry accounting, the comparison concedes that a library full of books remains worthless as long as it does not maintain a single book about these books. Only a catalog allows specific access to the stored knowledge that can produce profit by way of reading. This insight was later taken up by another famous counselor and writer in Weimar, namely, Goethe.

Catalogs are not only rare in Baroque libraries—where they appear, they usually resemble the form of what they record: bound books, bound catalogs. Besides a list of the directories to be compiled (inventories of books, paintings, "curiosities," "index nominalis, [...] index materiarum [...] index librorum historicorum [...] conspectus materiarum [...]"), Leibniz's proposals for an indispensable library guide that mark the beginning of his activity in Wolfenbüttel in December 1690 include ideas on the form of cataloging: "paper slips of all books, sorted pro materia et autoribus."[57] The plan anticipates registering every book merely once, precisely on a slip of paper, so that the slip only has to be placed in the right order for any catalog organized alphabetically, by subject, or in any other way. Theoretically, this procedure could have successfully made numerous catalogs with the same data set. However, the plan is never carried out. In fact, the librarians supervised by Leibniz manage merely to assemble an alphabetical catalog; all the other plans fail for lack of employees and funding, a result that must have seemed only too familiar to Leibniz in the Projecting Age—as Daniel Defoe describes the end of the seventeenth century.[58]

In practice, the procedure turns out to be fairly easy. Walking past the shelves, the librarian makes copies of titles. "I will subsequently call them paper slips for short."⁵⁹ With meticulous exactitude, the librarian records extensive details of the work on the paper slip, as with the Josephinian card catalog of the Vienna court library (see next chapter): "1. Case number, 2. Author's name, 3. Format of the book, 4. Commentators, translator, and editor, 5. Year and place of printing, 6. [...] shelf number, 7. Format and number of volumes."⁶⁰ Moreover, "the title of the book must be copied neither too briefly, nor too extensively. It must be just so you can get a hint of the content of the book from it."⁶¹ Thanks to internal mobility, or the permanent potential for reordering, the index catalog emancipates the order of the library from its physical shelving locations. From then on, the scholarly catalog works and prevails. Its paper slip form in Wolfenbüttel in Leibniz's time nevertheless remains only one temporary aid, again gathered and bound into the printed catalogs that librarians value more highly than fleeting slips of paper. Only an accident, as the next chapter will demonstrate in detail, will unexpectedly turn this procedure into a permanent order.

What remains pivotal, however, is the relation between index and book, which implies both temporary and permanent cataloging. For in contrast to the fixed entries of a continuous list on sequentially linear pages, paper slips can be reconfigured as freely mobile units in ever new arrangements. A slip of paper serves as a first pointer, which refers with the help of a call number to an address, the place the text occupies on a shelf. However, it not only points to the location where a text is found—it also embodies a highly compressed data set that characterizes the book to be found. Ideally, the slip of paper contains not only complete bibliographical specifications (with detailed title, subtitle, authors, etc.), but also a short content key. Thus, it delivers a derivative of the text it represents. More than a mere administrator of access, the "title copy" becomes a representation of the text—which now need no longer be read every time. The representational feature deflects from what it refers to; reading the table of contents—or even more briefly, just the title—protects against having to read further. Thus, it is barely surprising that around 1800, this form of representation on mobile paper slips experiences a first boom. An analogous situation is the use of visiting cards: "One arrives at one of the famous spas, a couple of hours after arriving one sends out a few hundred visiting cards, and the same day one is introduced to the whole society of the resort, and acquainted with two to three hundred people as if one had already lived with them for many years."[62] Making acquaintances via visiting cards substitutes for personal meetings—something that has become routine in our virtual worlds.

Returning from the drawing room to the library, we can state for the time being that its "wealth" can be found only if it is also registered in the catalog. It is hardly accidental that Goethe, entrusted with supervising the Jena and Weimar book collections, describes his impression of the Göttingen library in economic terms: "One feels as if in the presence of capital that noiselessly yields unpredictable interest."[63] But what does this wealth consist of? "The capital is the mass of writing accumulated in text processing, and the yield is texts that originate from the loops of the bureaucratic-literary processing."[64] At the very beginning of this paper machine that eventually produces novels and learned texts stand the anonymous catalogs without which the material is inaccessible.[65] The slips of paper in the catalogs become a derivative of the registered writings, the interest rate of amassed capital. The higher the magnitude of indexing, the greater the later yields, in the form of ever new texts resulting from texts thus made accessible. The library becomes a bearer of capital, a data bank, lending information as if it were credit. The latter is reliably paid back in the indexing of new writings, whose contents in turn feed on the old ones.

In 1716, Leibniz dies and the library at Wolfenbüttel falls into neglect, catalogs are carelessly kept, and the probability of discovering a text with the help of the book catalogs dwindles. In the course of the eighteenth century, finding anything in a library that has inadequate catalogs remains a challenge—not just in Wolfenbüttel, but everywhere else as well. To the extent that the bibliographies of Konrad Gessner fulfill their mission and lead to the production of new books, the complaint about a flood of new books is strengthened. The next chapter outlines what kinds of canalization technologies provided relief to the libraries of the eighteenth century as they were being flooded with new printed matter.

I Around 1800

3 The First Card Index?

Addressing Ideas

You may say that it isn't necessary to read every last book. Well, it's also true that in war you don't have to kill every last soldier, but we still need every one of them.
—General Stumm von Bordwehr, in Robert Musil, *The Man without Qualities*

"You haven't got lost, have you?" asked the usher in amazement, "you go down this way to the corner, then right down the corridor straight ahead as far as the door." "Come with me," said K., "show me the way, I'll miss it, there are so many different ways here." "It's the only way there is," said the usher.
—Franz Kafka, *The Trial*

On Christmas Eve, 1770, a court decree by Her Majesty Maria Theresa goes out to the mayor of Vienna, ordering him to "make the numbers on all houses legible and visible, on punishment of 9fl."[1] This refers to the so-called *conscription numbers* that serve to simplify the registration of the male population of Vienna so as to include all possible conscripts. Yet since "difficulties" crop up in carrying out this administrative act, the mayor obeys the royal order in trying to create transparency and asks the town council in early January 1771, after Epiphany, to mark houses in the course of the "general conscription of souls and houses" not only on the outside, but on the inside as well.[2] The purpose of this Christmas gift from on high aims at remedying a perilous military shortage, drafting every potential soldier on the recruitment lists without exception. It is not so much a matter of being able to deliver a draft summons *on time*, but of the delivery *itself*.

Before the appearance of the first house numbers, the only thing serving as orientation in the warren of Viennese streets for the administrators of the conscription catalogs—who in pursuit of their duties draw up one of the very first card catalogs—is a purely empirical knowledge of the neighborhoods, or the equally empirical and reliable questioning of locals.[3] What Friedrich Adolf Ebert praises in 1820 as "local memory" is the only factor that enables navigation through the town's anonymous lanes.[4] Owing to the lack of a more thorough addressing logic, places are accessed thanks to the systematic arrangement of laborers—that is, the crafts and guilds distributed throughout the neighborhood.

Until an address system (directed only at men) is set up in early 1771, with conscription numbers for every single house, inside and outside, the indistinct *sea* of houses offers convenient hiding places for delinquents and deserters.[5] Therefore, anyone obstructing governmental access to men liable for military service by "making house numbers illegally unreadable or invisible" is threatened with arbitrary financial or bodily punishment.[6] To prevent this sort of crime, rewards are offered, funded from the fines raised in this manner, as an incentive to denounce deserters.

By 1777, the government of Lower Austria starts a renewed numbering of houses. "As many new houses were built after the last conscription which have no number yet, this is also an opportunity for the rectification of the house numbers." New entries are to be treated as follows: "If for instance three new houses are found between numbers 12 and 13, the first is to be 12a, the second 12b, the third 12c."[7] Moreover, the conscription decree further increases the depth of addressing, including "women, Jews, and farm animals."[8]

New difficulties with the exactness of addressing appear within a decade after the second successful numbering in 1777. Although houses are marked on the inside and outside, the neighborhoods themselves still present mnemotechnical and navigational challenges. There is still no unequivocal naming of every lane, street and public thoroughfare. "Many large and small lanes, their names unknown to most, force not only strangers, but even locals unfamiliar with the area to search or ask for directions, their wandering about causing such embarrassment and delay that the order has been passed that street names be written in bold black letters on the first house at each end of every lane, large and small lanes alike, in a similar fashion as pertains to the military conscription numbers."[9] This unambiguously fixes the locations of (and thus the state's access to) military conscripts, women, Jews, and cattle. In wartime as well as in the

The First Card Index?

library, as General Stumm von Bordwehr attests, units are subject to fluctuation. "The purpose of the naming is unequivocal and urgent: to guarantee quick retrieval and return of a used book, and then to confirm its being available in an inventory audit."[10] Books and recruits must be registered incessantly as new entries, and just as incessantly as war losses, so they can be exactly and reliably addressed. "Thus the call number issues from its location."[11] Just as the conscription number is aimed at establishing the possible location of a respective recruit, so the call number of a book ideally carries the address of its regular storage place. Hence, it is hardly surprising that more recent library management uses the postal analogy of efficient addressing when it attempts to circumscribe the heterogeneous and often contradictory history of the development of call number systems.[12]

One fundamental problem with localized, shelving-based unit call numbers (which in turn carry authors' marks) inevitably arises when massive new acquisitions occur or buildings are built.[13] A book flood exceeds the capacity of even the largest repositories. Even if there is no acute shortage of space, librarians are forced to engage in time-consuming sorting and rearranging. Besides, those in command worry that "as a result of frequent handling by subordinates, books can easily be damaged."[14] To diminish these occasionally unavoidable disturbances to order and to limit access to trained librarians who can provide readers with the desired books, a new type of mobile management becomes necessary in turn: reference management that finds books no longer via their location on the shelves, but via their systemic place in an alphabetical or subject catalog.

This change in the search procedure, from the level of books in a repository to the level of order in catalogs, proceeds rather slowly. However, what happens immediately is that the shift from tracking books on shelves to searching for their representation by bibliographical title copies also marks a shift in library order, a fundamental shift in its inner structure and architecture, and in directors' plans for the logistical design of book storage.[15] It seems to be the fate of libraries that a particular order always coincides with a director's term of service. As soon as a new director, prefect, or manager takes over, one of the first acts tends to be rejection of the present order in favor of establishing a new, often completely different one, mostly legitimized by the allegedly encountered chaos that almost forces reorganization. "I have not yet read the biography of a librarian which does not report that he found the library in complete or partial disorder upon taking office.

Then is recounted all that he did [...]—and were we to receive his successor's biography, it would in most cases begin exactly like his predecessor's."[16] Ad infinitum? More pivotal than the inevitable reorganization is the step that replaces subject shelving with the systematic representation of books in a complete catalog. Library apprenticeships and library histories tend to sum up this disparate development rather laconically. "Over time, people gradually ceased using a *fixed* system that places every single book on a specific shelf whose name it bears for good, and moved to a *mobile* system."[17]

Precise addressing remains a necessary condition of flexible data management for registering recruits or books in situ. At best, conscription numbers and call numbers register the exact location of the address—but at the very least, this is gleaned from the town map or library plan. At the same time, exact addressing empowers a higher logic for representing inserts: new entries are incorporated into the regular order of lists, either according to scientific criteria or alphabetically. Instead of merely being inserted, as before, between existing units, they may safely be added in the order of their appearance: a running number procedure (*numerus currens*) in the library as the librarian's answer to the flood of books. "A book's location is extremely unimportant," as the farsighted library-analyst Albrecht Christoph Kayser states.[18] Hence, the question of where to find which book is no longer directed toward a particular shelf; rather, it is directed to the symbolic order of the catalog. Thus, the need for a mobile (and adaptable) systems is moved to the catalog itself. "To achieve the purpose of the call number—namely, to prevent misplacing a book—books must be equipped on the inside and the outside with the individual address, which is also noted in the catalog."[19] Provided these basic rules of addressing receive proper attention, the catalog may show the way to the desired book as a local guide directs visitors to the correct door.

This is how mechanical shelving replaces local shelving call numbers, positioning books by their *numerus currens* on the shelf. In the Viennese university library, reopened in 1777, instructions for arranging the "treasury of knowledge" (Leibniz) advise installing books according to a "systematic plan of the sciences, and consequently according to the future library sections," so that every book can be found by means of the code *Roman numeral / Roman letter / Arabic numeral* (for example XIV.B.12).[20] Nonetheless, one has to make sure "not to place books too close together, for even a small increase in their number would necessitate substantial transfers."[21] Despite considerable dispersion, the estimated free space is soon filled in by growth in the collection, so that the next iteration of the library rules in 1825—faced with a lack of discipline and systematic care on the part of librarians—abandons subject shelving even more decidedly. Essentially, it follows its predecessor of 1778, rendering the catalog the only search instrument.[22] Paragraph 22 reads, "Books are lined up according to every main subject or every main class and in every section according to format and size. It follows that no further scientific organization is possible at the shelves, so searching must be done not in the display of books but in the classified catalog."[23] Six paragraphs later, an additional call number procedure is introduced, in the form of "consecutive numbering throughout the entire library."[24] This procedure, "by which all the library catalogs are completely disconnected from and rendered independent of their local order and local call numbers," explicitly refers back to Martin Schrettinger's (1808) *Outline for a Textbook of Library Science*, which was rather controversial in 1825. One consequence of the new library science is that from then on, individuals no longer need to store contents; they can simply read addresses.[25] For the *header* of these addresses not to fail like the old order if the librarian falls off a ladder, these addresses need to be noted in the catalog in lieu of a local memory[26]—another consequence of Schrettinger's library science. For it is "a common mistake of employees that they believe they will live forever, and thus arrange their shops without regard for their successors, considering them well kept in their own memory, without written notes, thus making it impossible for those eventually taking their place, or at least making it infinitely more complicated for anyone to pick up the thread of those who have been called to meet their Maker."[27]

This is not to stretch the analogy between house numbers and book call numbers since the late fifteenth century,[28] but rather to point to a coinciding addressing logic during the same decade and in the same town, both guided by the same phenomenon: a reaction to *mobility*. Both forms of addressing are designed to account for units that threaten to disappear among countless masses. In addition, both are united in the logic of a list whose entries register the respective depository alongside the names of those "deposited"—the authors or recruits. The numbers applied on Viennese buildings create the possibility of an address in the sea of houses, and thus a grid can be drawn for the exact whereabouts of recruits. By the same token, the rising book flood forces a gradual turn away from fixed shelving order and toward more delocalized addressing. It is not sufficient to furnish the library with catalogs by decree. Rather, detailed and exact written procedures are needed to guarantee the logistical architecture of the library beyond the fluctuations of a term of office. Faced with data streams whose sources make mobility necessary, the logic of search progresses from an approximate system—walking along the shelves (as if in a parade of books)—to specific access via the catalog.

Data Streams

The pattern from which the *catalog movement* (the increasing manufacture of catalogs as search tools) departs in the eighteenth century is not so much the highly praised and defended catalog, the *book* of books. Rather, at this time it is common to *walk* though the subject shelving installation paradigmatically, like a human search engine, as it were. Gotthold Ephraim Lessing, deeply in debt and driven to Wolfenbüttel as a refuge from his gambling addiction, discovers the books in his library without a catalog by walking.[29] On extended walks and accidental turns through the halls and past the cabinets of the library, he discovers forgotten and unexpected books. As a result of these constant rambles through the stacks, Lessing neglects his librarian's duties, and not only those of cataloging.[30] Instead, he finds exceptional works for whose discovery posterity still praises him today. Without the support of a register, without the aid of overviews, his discoveries are the product of random access.

However, the narrow coupling of shelving and Enlightenment (which demands an academically differentiated access to the knowledge of the library) reaches a high degree of complexity with the classification system sketched by Gottfried Wilhelm Leibniz and his 1680 proposal for a universal library—not to neglect influential precursors in the scholarly system, including Gabriel Naudé.[31] In contrast to the philosophical encyclopedic systems ruling at that time, he recommends shelving books according to systematic concepts, ordered by academic fields and arranged according to current interests.[32]

There are numerous competing systems in the eighteenth century, although few designs take the diversifying of the sciences into account, thus proving incapable of integrating new branches of knowledge. For this reason, and owing to the library practice of unrecorded insertion on the shelf, the system in the end prevents direct access to the books, thus depriving users bit by bit of the knowledge of its treasures. It was considered sufficient for a new entry simply to be placed on the right shelf according to its supposed content, often without recording its title in the catalog—as the book was expected to be found there on the shelf. "Hence, for a long time many librarians held the view that catalogs were not necessary at all; and until the nineteenth century library owners also believed that catalogs were the private projects of librarians."[33] Thus, the access routines necessarily culminate in Lessing's arbitrary celebrations of a *random access memory*. "Dictatorial power of the catalogs over books" takes hold only at the end of the age of Enlightenment, at the beginning of the Napoleonic age, and as a consequence of the French Revolution.[34]

While Lessing, between 1770 and his death in 1781, does his best to refrain from card games and other gambling activities by strolling through the Wolfenbüttel book collections, playing cards find a different application in the cradle of the Enlightenment. In 1775, the Académie des Sciences in Paris appoints Abbé François Rozier to tabulate an index of everything the academy published between 1666 and 1770 and to draw up a general index. Rozier chances upon the labor-saving idea of producing catalogs according to Gessner's procedures—that is, transferring titles onto one side of a piece of paper before copying them into tabular form. Yet he optimizes this process by dint of a small refinement, with regard to the paper itself: instead of copying data onto specially cut octavo sheets, he uses uniformly and precisely cut paper whose ordinary purpose obeys the contingent pleasure of being shuffled, ordered, and exchanged: "cartes à jouer."[35] In sticking strictly to the playing card sizes available in prerevolutionary France (either 83 × 43 mm or 70 × 43 mm), Rozier cast his bibliographical specifications into a standardized and therefore easily handled format. Playing cards offer numerous advantages: only after 1816 do their hitherto unmarked backs (figure 3.1) assumed a Tarot pattern. Therefore, they were frequently used as lottery tickets, marriage and death announcements, notepads, or business cards.[36] Besides their widespread availability, their uniform measurement allowed easy shuffling. Their relative stability (in comparison to sheets of paper) afforded robust handling. Rozier is so convinced of his procedure that he recommends that readers treat the academy's future publications in the same way—that is, by collecting them on playing cards that allow them to be sorted by using the unprinted flipsides.

Figure 3.1
French prerevolutionary playing cards. (From François 1974, p. 81.)

However, while the idea of *shelving as Enlightenment* is lost in playing cards[37] in France with revolutionary momentum, Germany, thanks to an influential advocate, holds out for quite some time, so that systematic shelving remains in effect until the start of the twentieth century.[38] What enables the nineteenth century to manage its steadily swelling book flood, so that one can find every single volume in it? Here we need to look more closely at some of the prerevolutionary processes. Whereas in Germany, the subject parading of shelves remains the general procedure, Vienna around 1780 quietly and almost *unconsciously* begins to experiment with an independent practice that eventually will help establish the card index as the one and only library search engine.

Copy Error: The Josephinian Card Index

"Work has started on May 22, 1780," reads an addendum to the *Rules for the Copying of All Books in the Royal and Imperial Court Library*[39] by the prefect of the library, Gottfried van Swieten. The project consists of preparing a complete catalog of the court library including its entire printed inventory.[40] In the Vienna court library, there had not been a complete catalog since Blotius's time as head librarian. The collection is in desolate shape as late as June 25, 1745, when Gerhard van Swieten (Gottfried's father) receives the prefecture of the court library (offered shortly after his appointment as personal physician to Maria Theresa), as a sign of Her Majesty's gratitude for his service.[41] "Thousands of books lay unbound and consequently dead for use."[42] In 1766, under Gerhard van Swieten's prefecture, numerous catalogs (each registering a separate section that itself had again been dispersed) are combined.[43] This merges the "aggregates" into one comprehensive catalog.[44] The reference book is arranged alphabetically by author, and in 1772—shortly before Gerhard van Swieten's death—it comes to a temporary conclusion with the seventeenth volume. In spite of several supplementary volumes, it remains uncertain whether the promise of a *universal* catalog was fulfilled; perhaps it ought to have been called a *provisional* catalog instead. The entries are too varied for a standardized description of the books. Nonetheless, the universal repertory suffices as an interim solution until, starting in 1780, the Josephinian catalog begins to describe the books on the shelves. Nonetheless, given the state of the library in 1745, the universal inventory rightly counts as one of Gerhard van Swieten's most extraordinary achievements during his 27-year term of office as prefect of the Vienna court library.[45]

Floods

While administrators in the Middle Ages anxiously noted library *growth* as actually an "inventory decline" after the annual checking of their inventory lists, the incessantly rising data flood of printed works provides library historians with a new criterion for assessing library management.[46] The German-language book market experiences a noteworthy increase in productivity during the term of the van Swietens' Vienna court library project.[47] Yet before we investigate the catalog as the canalization of a tidal wave of books, we should ask what the inventories of the court library after 1750 are actually made up of.

Between 1760 and 1765, printing begins to accelerate rapidly and monotonically (were it not for declines during the Napoleonic wars, one could say "strictly monotonically").[48] Numerous innovations in book production and literacy increase the reading audience, and the new demand for literature changes reading habits.[49] Apart from this unexpected demand for literary products, a consequential administrative act precedes the rising tide of publishing, allowing for steady growth, at least in Austria, of deposit copies,[50] reinforced by the occasional annexation of independent book collections.

In 1773, Pope Clement XIV declares the Jesuit order dissolved in Austria and its hereditary lands. This decision was preceded by decades of furious struggles, starting in 1713 in the Austrian Netherlands with the persecution of Jansenists that aimed to break the notorious political influence of the Jesuits.[51] In Austria, this struggle intensifies after Gerhard van Swieten is appointed personal physician to Maria Theresa and entrusted with the complete reform of the Austrian educational system. His educational reforms include jurisprudence and his own field of medicine[52]—and finally, by means of decisive actions, he gives every faculty that had previously been in Jesuit hands (philosophy, theology, as well as various administrative offices) over to Jansenist control.[53]

Gottfried van Swieten, perhaps biased against the Jesuits because of his own strict education at the Vienna *Theresianum*, continues his father's reforms—above all his struggle against the *Societas Jesu*.[54] On December 6, 1779, he sends out a decree: "After Our Majesty, Our Most Gracious Lady, learned that in the course of the dissolution of the Jesuit order important papers were found regarding the legal estate and other circumstances of this former organization, Your Excellency is reminded in Her name that there should be a catalog of all these writings and documents, an exact and dependable directory compiled as soon as possible by me."[55] This order to surrender all printed matter is soon followed by a selection process led by Gottfried van Swieten himself, adding to the court library everything that had not been included previously, with the aid of the Jesuit catalogs.[56] After the arrival of catalogs listing inventory that had not been intentionally destroyed, van Swieten makes selections—as he would have done as the court commission censor, yet here guided by the differentiation not between *forbidden* and *permitted*, but between *valuable* and *worthless*, in the service of a more general, transgenerational, conservationist consciousness.

Gottfried van Swieten's term of office is also marked by another short, yet powerful influx. If the integration of independent collections (such as, in 1761, the private library of Maria Theresa's father, Charles VI, and other ducal collections) had at times brought substantial growth in terms of quality *and* quantity to the Vienna court library, the friendly acquisition of the city library of Vienna on July 31, 1780, constituted another important source.[57] That collection alone contained 1,182 volumes that the court library did not yet possess, as well as 1,215 editions different from those already there. The books were labeled "former city library Vienna" and integrated into the inventory.[58] Around the same time, the Josephinian catalog project commences, which over the subsequent months will encompass a description of the entire inventory of the court library's shelves.

In 1750, Maria Theresa reminds publisher Johann Thomas Trattner "that it is our state policy to have books produced, we have almost nothing, so many must be printed."[59] However, the shortage is not remedied the way Maria Theresa intended. "Reprints must be made until the original editions can be acquired. Proceed. Sonnenfels will instruct you as to which ones!" By the end of her reign, the dissolved monastery libraries begin to flow toward Vienna, plus volumes from additional closures of Jesuit communities in Krain, Tyrol, and Styria under Joseph II's rule, this wave subsiding only in 1787.[60] Volumes superfluous to the Vienna court library are used as payment in exchange for other, more valuable works.[61] Eliminated and banned books can still pave the way to other titles. "Reports of the destruction of books in transport from Lilienfeld in 1789 are well known. At that time, books were used by the coachmen to patch bad roads."[62] Thus, the influx decreases, and yet the court library overflows.

Canals

In the midst of the library's impending collapse under the superfluity of books, we now need to look at the librarians' reaction and cataloging techniques in response to the influx of books. What kind of canalization can cope with the data flow pouring in from the former monasteries? Four main administrative tactics are put into effect.

The successive comparisons between the cloister inventories and the court library's stock serve as a first filter. This evidently requires a complete catalog of existing books, not least for comparison's sake. The insufficient usability of the universal inventory of 1766/72 must have become obvious at this point. Any volumes that turn out to be duplicates or rejects are assigned to provincial universities.[63]

The second tactic consists of trying to remedy the problem at the root—by setting up a quality filter to prevent the publication of "pointless brochures."[64] Counter to Maria Theresa's order of 1750 (to overcome the shortage of books by reprinting), Gottfried van Swieten feels impelled to take measures against a flood of academic publications that is more or less the direct result of his own educational reforms.[65] By April 1784, authors must deposit a sum of money that is refunded only if the court agrees to the printing; otherwise, the amount is withheld and turned over to welfare.[66]

Censorship is effective not only in the struggle against the Jesuits, but also in regulating the book flood.[67] As a third tactic, Gerhard van Swieten keeps adding to the Roman Catholic *Index Librorum Prohibitorum*, first published in 1559, appending to the Austrian secular version his brief judgments on numerous additions.[68] While Gerhard van Swieten concentrated on scholarly texts, his son Gottfried shifts horizons: "Seen from the most innocent standpoint, Cabalistics and magic are the fruits of a weak and unwholesome mind and must be relegated to the realm of insanity."[69] Furthermore, he protects the court library against theological and legal writings; a significant portion is turned over to the military for use in manufacturing ammunition.[70]

The most important mechanism available to the librarian for countering the press of new knowledge via librarianship is indeed the catalog itself, which cannot avoid registering the additions, taking the full brunt of the flood. Drainage is engineered—that is, the court library is conceived as a central storage tank. "The intention of the court order of Her Royal and Imperial Apostolic Majesty was to collect the legal estate of the Jesuits dispersed everywhere in documents, books, and writings, and to bring them to a place where they would be preserved, ordered, and consulted. Such a place is indisputably the court library, the only one in the country with which a collection of this kind can be combined."[72]

However, until work on a new complete catalog begins in May 1780, new entries cannot reach their systematic place. Owing to a lack of manpower, the influx is met with a typical administrative reaction. Gerhard van Swieten already "noted upon making closer acquaintance with the library that the care and generosity with which the Austrian rulers sought out and purchased the most exquisite books in the whole of Europe was far greater than the diligence of the prefects and custodians in ordering and preserving these acquisitions."[73]

The Algorithm

The undertaking that begins on May 22, 1780, later to be called the *Josephinian catalog*, is extant in "205 small boxes" in an airtight locker in the

The First Card Index? 39

Austrian National Library; it is widely, and often proudly, considered the first card catalog in library history.[74] Before we ask whether this is accurate, let us compare the process with Konrad Gessner's method. Three aspects distinguish the Josephinian catalog from the paper slip techniques practiced since early modern times: written instructions for the cataloger, a division of labor organized around interfaces, and the duration of catalog use. Only the coincidence of these three characteristic features and their mutual dependence separate the enterprise specifically from prior attempts, such as the refinement of the Wolfenbüttel catalog under Leibniz, or Rozier's excellent index based on playing cards.

Because of the constantly growing number of volumes, and to minimize coordination issues, Gottfried van Swieten emphasizes a set of instructions for registering all the books of the court library. Written instructions are by no means common prior to the end of the eighteenth century. Until then, cataloging takes place under the supervision of a librarian who instructs scriptors orally, pointing out problems and corrections as everyone goes along.[75]

The instruction is elaborated in two phases. The first design is van Swieten's own *Instruction and Guidance for Those Who Copy Titles and Books*.[76] Besides a list of assistants, these "rules for the copying of all books in the Royal and Imperial Court Library" encompasses seven pointers on what is to be noted along with the description of books on slips of paper.[77]

Adam Bartsch, the fifth scriptor (and later head of the Print Collection), refines these instructions further, complementing the necessary bibliographical requirements with a set of detailed and anticipatory behavioral guidelines.[78] These commands are further enriched by a flowchart, guiding order and approach to coordinated labors. The final version is sent to van Swieten to serve as a standardization program. The goal of a unified, adequate catalog can be reached only if one can ward off the risk that the data will be arbitrarily diversified whenever employees act randomly.

These are some remarks that might find room in a further edition of the regulations for furnishing a new catalog, at least in part, so that I allow myself most obediently to send them to Your Grace. Now it is up to you to check and to use them. A similar regulation, which every copyist employed in the catalog project should always have at hand, should not merely instruct on how to write a catalog, or the single components of the same, but serve as a guideline that indicates that everyone should work in the same way. I assume at least that each of us knows how to note down book titles, yet the differences in how they might be written, even if each was itself without error, in the end would produce an inconsistency on the whole that would make the catalog if not unclear, then at least deprived of adornment and the proper appearance, being designed for one purpose, from one point of view, and according to one system only.[79]

Thus, in a reversal of the hierarchy, the plan of the plan goes back to His Baronial Grace, to be initialed and countersigned: the instructions return without alterations to their author Adam Bartsch—this time, however, he is the addressee of the assignment. Yet a brief look at the executives and staff of the court library in spring 1780 casts doubt on whether there is sufficient manpower. Besides prefect Gottfried van Swieten and his director, the court library employs two curators, five scriptors, and four library servants.[80] Van Swieten hires seven additional catalog assistants; with this team, the court library has attained a peak employment that, not counting the usual fluctuations through attrition and hiring, it will not reach again until the late nineteenth century.[81]

Writing heads and reading heads are instructed by Bartsch's program to expect command input from His Baronial Grace. The catalog plan is codified, compiled in library technology, and linked to the library function.[82] However, the data storage of the court library's mass memory has yet to be called up. To satisfy the architecture of a modern process, a space separate from the usual library business is furnished, a catalog room or working memory for a central bibliographic unit. In this CBU, the program processes data contributed by various paths. Yet this data flow only joins run-time when the transfer from mass storage into processor memory begins. Then the processing can start: from the library, servants fetch the books shelf by shelf; the scriptor checks the volumes for a position number using a Latin / Latin / Arabic code.[83] Any mistakes are corrected before the description process starts. Regardless of their position in the library hierarchy, scriptors, library assistants, messengers, and all other employees take the volumes and record formal categories such as title, author, place of printing, and year, format, printer's name, and any defects, putting them down on single uniform slips of paper.[84] Then the books are returned to their location in the spectacular library reading room, the *Prunksaal*. "Once all the books have been described in the accepted way, one can consider ordering and copying the slips of paper."[85] The next step of the

The First Card Index?

Figure 3.2
The catalog capsule. (From Petschar, Strouhal, and Zobernig 1999, p. 112.)

procedure tackles the normalization of specifications, especially different spellings of authors' names, before the slips of paper can finally be arranged in specially made catalog capsules (see figure 3.2).[86] An unambiguously defined interface remains a necessary precondition. Piled up, tied into packages, and bundled with a linen thread for further use, the slips of paper await alphabetical ordering.

The process of title storage and transfer onto fragmented cards advances rapidly: by the summer of 1780, 31,596 works in 27,709 volumes have been registered, and by the next summer, the remaining 23,434 titles are described completely. The performance of every scriptor is accounted for in a "list of the number of books described in summer 1780."[87] Adam Bartsch, author of the instructions, chalks up the biggest contribution according to this interim balance, with 4,637 works in 5,372 volumes. In the end, the Josephinian catalog represents the entire library on approximately 300,000 cards, including a distinctive reference system—successfully meeting the first partial goal of registering all titles.

Error: Buffer Overflow

The further cataloging plan anticipates using the standardized and sorted paper slips as a basis for bound alphabetical and subject catalogs. "From just these slips of paper, which are to be carefully stored, and to which one must apply the utmost care so that, being cut up, they are not scattered, one could, in time, generate a subject catalog without having to undertake the laborious work of a renewed description of the library's books."[88] However, work on a bound catalog never begins. Despite a deep-seated distrust for the loose order of paper slips that threaten to flutter about with every gust of wind,[89] the court library staff does not copy the loose elements into the paradigmatic bound catalog in book form.

The question remains of how long it would take to incorporate every library slip made since Konrad Gessner. How much time must pass to abolish the stigma of unavoidable loss of order before the new status of fluid harmony might be reached? A merely interim solution unexpectedly gets a chance to prove itself, before the index card is recognized and respected enough to establish itself in the catalog room. By 1912, when the struggle over catalog forms is by no means won, Fritz Milkau, an opponent of the indexical representation of books in the catalog and an ardent advocate of subject-classified shelving, denounces the apparatus of the card index as an "embarrassment."[90]

Around 1780, nothing points to the fact that the Josephinian catalog will be implemented as an actual search engine or storage representation. A cautious surmise by Gottfried van Swieten, in his general report of 1787, indicates warily that it is not feasible to catalog the oppressive mass of books—in the shape of either an alphabetical or a subject catalog. In his report on the catalog project from 1780/1781, he reflects on it as "an enterprise of major dimensions and effort—the materials alone fill 205 small boxes which, once processed, will bring forth a catalog of an estimated 50 to 60 folio volumes."[91] The inability to carry out this ambitious plan results in the extended temporary use of the index cards. Instead of producing the book of books, the Bible of all libraries, as prescribed by every routine and library practice to date, one boldly keeps the bibliography divided into discrete miniature bibliographies. Because the abundance of books prevents the completion of bound catalogs, a preliminary, interim paper slip solution finally gives birth to the card index. Indeed, the Josephinian card index owes its continued use to the failure to achieve a bound

The First Card Index? 43

catalog, until a successor card catalog comes along in 1848. Only the absence of a bound repertory allows the paper slip aggregate to answer all inquiries about a book's whereabouts after 1781. Thus, a failed undertaking tacitly turns into a success story.

Paper Flow: Taming, Duration

While the Josephinian catalog goes into production in 1780 and eventually into permanent use, the nearby Vienna University Library continues to follow a general direction that should not go unnoticed in our investigation of the primal scenes of the independent card index. In 1756, the holdings of the 391-year-old university library had been incorporated into the collections of the court library. In the course of the Josephinian reforms, this takeover caused perceptible reductions in the university's book use, as it had been more interested in the circulation of its books than in treating them as museum pieces. As a result of these deficits, the university library regains independence; however, when it celebrates its reopening on May 13, 1777, it has not recovered the books lost to the court library. The foundation of its new collection in the service of science will be the Jesuit book flood:

Decree of the Education Commission at Court, March 24, 1775
informing the government of Lower Austria that Her Majesty grants the local university use of the libraries located in three closed Viennese Jesuit colleges, for whose swift arrangement and incorporation into the *Collegio academico* the government must bear the necessary expenses.[92]

This influx increases once again after 1782—though lagging behind the court library, which retains the prerogative of first choice—as "the university library of Vienna [receives] from the discontinued cloisters of Lower Austria another influx that proves fruitful, also by providing numerous marketable duplicates."[93]

Entrusted with equipping the university library, the director of the theological faculty, Benedictine abbot Franz Stephan Rautenstrauch (1734–1785),[94] devotes himself to the production of a "Catalogo Topographico," following the principle of using it as a basis for all other catalogs, "composed only of tiny slips of paper that could afterward be transferred as needed to any other catalog."[95] Rautenstrauch derives his bibliographical suggestions—down to passages copied verbatim—from the *Introduction to*

Bibliology by Michael Denis, who joined the court library in 1784.[96] Since Denis makes no special mention of paper slips, the honor of founding the procedure of cataloging on paper slips in the sense of Blotius and Leibniz belongs to Rautenstrauch. The first three points of his instruction set concern the order of books in the library space,[97] followed by six points on cataloging, of which points 4 and 5 pertain to paper slips:

4. What follows is the compilation of the basic catalog; that is, all book titles are copied on a piece of paper (whose *pagina aversa* must remain blank) according to a specific order, so that together with the title of every book and the name of the author, the place, year, and format of the printing, the volume, and the place of the same in the library is marked.

5. This catalog is then to be cut up into single slips of paper so that every slip records only one book. These slips of paper are then separated in two parts, one with books and authors, the other with anonymous works, and then both are compiled according to alphabetical order and copied according to samples A and B, and joined together to create an alphabetical catalog [see table 3.1].[98]

Given what we know about bibliographical practice since Blotius's work in Vienna, this approach is not surprising— nonetheless it is two years before the characteristics of the Josephinian catalog become clear, it remains extraordinary. While Adam Bartsch's instructions apply only to the Josephinian catalog project, Rautenstrauch's regulations are free of the idiosyncrasies of the court library and can apply to all university and school libraries in Austria. Thus, they become an imperial standard for catalogs, over thirteen years before the regulation generally cited as the first of its kind, whose rules for description have more than local validity, extending to all parts of France: the efforts to create a French national bibliography in 1791.[99] While index cards in the court library unexpectedly turn into permanent equipment, the librarians of the university and elsewhere faithfully copy the order imposed by the basic catalog into an alphabetically ordered, bound catalog. In the case of the Viennese university library the result, including supplementary volumes, encompasses the entire inventory by 1810.[100] In the course of this thirty-year project, the fundamental paper slip principle continually proves itself feasible, so that finally, the basic catalog is not destroyed, but maintained.[101] Abbot Franz Stephan Rautenstrauch, however, succumbs to gastric fever in Hungary—a sudden death, perhaps even murder committed by Jesuits, as some of his friends speculate.[102]

The First Card Index? 45

In 1823, Johann Wilhelm Ridler finally declares the card index the indispensable basis for cataloging. Ridler complains that the paper slips were "disorderly and in a state of decay due to an alarming carelessness; some were even used to the foulest ends by library officials."[103] Hence, it does not come as a surprise that Ridler takes preventive measures for his newly erected basic catalog.[104] Under this new order of things, the paper slips are accumulated as representatives of the floods of books to guarantee long-term access. The flood is managed not merely by bound catalogs, but also by the card catalog. Even in the twenty-first century, Viennese libraries record every new entry published before 1931 on a clean sheet of paper, in that same basic catalog from 1823—optimized only by an audit in 1884.[105]

Revolution on Playing Cards

A lot disappears with the departure of the ancien régime, and yet Abbé Rozier's idea gains entrance into revolutionary legislation with playful ease. The French book flood crests on November 2, 1789, after the National Assembly decides to declare all clerical goods national property, including the book collections of the church.[106] These confiscations require finding a way to tell the nation what belongs to the nation. General stocktaking begins in 1790, stopping a sell-off of the books—mostly for the mere price of paper—and keeping them in public libraries. On October 16, 1790, a committee is formed that devotes itself twice a week exclusively to the question of the mobile goods of the church, particularly what to do with books and art. The result appears in the form of a set of instructions on how library catalogs should be generated from now on.[107]

As proposed by one of its members, Louis-François-de-Paul Lefèvre d'Ormesson, the committee pursues the idea of compiling a national bibliography of all French books on the basis of inventory lists in the individual districts. A unified catalog could thus list the books with reference to local libraries to simplify their exchange and interlibrary loan, as well as the exchange of duplicates. Optimistically, three months are estimated for compiling the bibliography in Paris. Yet it soon becomes clear that this will not suffice, and it seems advisable to begin the procedure simultaneously in all eighty-three newly created *départements*. However, to be able to achieve this *multiprocessing*, the synchronous manufacturing of catalogs

Table 3.1
Catalogus autorum

Autorum nomina et Librorum tituli	Impressionis			Com-pactum	Cista et Series
	locus	annus	forma		
D: Ambrossi Mediol. Episc: Opera omnia	Parisiis	1549	Fol.	Franzb.	II. B

in the districts, the committee recognizes the need for detailed procedures with which to instruct inexperienced assistants. Furthermore, it remains essential to carry out the plan *in the same way* in all *départements* to guarantee completeness as well as to enable uniform processing of the catalogs delivered to Paris. Meanwhile, the first field tests with easy lists have demonstrated how little attention was paid to the collaboration by many local administrations. Requests to report on the local number of library holdings yielded only a moderate return, giving no reason to expect more from the project of a *national bibliography*: of eighty-three *départements*, only fifteen send correctly filled-in questionnaires back to Paris. Numerous administrations deny the existence of libraries, or claim they are worth little, while the majority prefer not to respond at all. Nonetheless, on April 11, 1791, the committee decides to compile a regulation for catalogers.[108] Less than a month later, on May 8, 1791, detailed instructions to teach untrained library employees how to compile catalogs *after an exactly agreed-upon procedure* are published. Explicit instructions point out that assistants are forbidden to attempt any reordering or reclassifications. Instead of dabbling in library hermeneutics, the rules prescribe a purely mechanical approach.[109]

In preparing these instructions, Gaspard-Michel LeBlond, one of their authors, urges the use of uniform media for registering titles, suggesting that "catalog materials are not difficult to assemble; it is sufficient to use playing cards [...] Whether one writes lengthwise or across the backs of cards, one should pick one way and stick with it to preserve uniformity."[110] Presumably LeBlond was familiar with the work of Abbé Rozier fifteen years earlier; it is unknown whether precisely cut cards had been used before Rozier. The activity of cutting up pages is often mentioned in prior descriptions. A full sheet of paper always served as a starting point, with the librarian gathering a number of titles and then cutting them apart by hand. Speculations about a shortage of paper as a consequence of the war, resulting in the use of (for example) playing cards, do not seem to apply to Rozier's pilot project.[111]

The work of the Bureau de Bibliographie, set up carefully in the Louvre, progresses rapidly and owing to the unexpected arrival of abundant new material is soon reinforced by more helpers. Numerous new employees are hired for an ever more differentiated division of labor, and soon enough, forty-three people are at work on the national bibliography. However, initial hopes for success flag over the course of sorting and processing—by 1794, at least 1.2 million cards record 3 million volumes. By no means all of the over 10 million works suspected to exist in the country ever reach Paris; only about 4 million are recorded on playing cards. Many *départements* either send slips with faulty specifications, or bound catalogs, or nothing at all.[112] In the end, not a single volume of the planned national bibliography is printed on the basis of playing cards. The records do not even see any use as an incomplete collection, as a catalog construction site. Despite careful instructions, the project fails because of disruptions in the data flow—the plan runs dry all too quickly. The procedure for bibliographical indexing itself again appears only as a step in a divided process, as in all previously described catalog projects since Gessner. The attempt to establish a lasting national monument on a truly contingent basis remains a purely temporary aid for the simplification of sorting.

Thus, in France the card index does not establish itself as a permanent search engine and user catalog before the mid twentieth century. However, as a catalog-technical innovation, the standardized form—that have been so happily reassembled centrally—undoubtedly takes hold. The switch to a uniform format for the description of all the nation's books, the change to agreed standards for cataloging on cards, remains a lasting heritage of the French Revolution. In that respect, the approach marks an impulse in the history of cataloging and the storage of knowledge. It is a matter not just of *égalité* instead of *diversité*, nor of the materiality of data carriers (paper versus cardboard), but above all of their *arrangement*: the insurgent paper is emancipated from the continuous ream and elevated to the precisely cut, standardized dimensions of the index card. In fact, the French Revolution brings about the transformation of the paper slip to the index card by virtue of material equality.

4 Thinking in Boxes

Nevertheless, almost everybody speaks better than they write. (This also applies to authors.) Writing is a highly formalized technique, demanding—in purely physiological terms—a peculiarly rigid posture. This corresponds with the high degree of social specialization it demands. Professional authors have always been inclined to compartmentalized thinking *(Kastendenken)*.
—Hans Magnus Enzensberger, "Constituents of a Theory of the Media"

When Konrad Gessner advocated in Zurich in 1548 that book indexes and descriptions should be made using excerpting tactics, he did so as a scholar addressing librarians as well as authors. The second half of the eighteenth century witnesses trends that help to dissolve this narrow coupling of librarian and scholar. Two lines of development emerge, differentiating formerly closely related functions. One points directly to the education of professional librarians, who regard the production of indexes as an inherent part of their occupation. A second path already gained considerable attention in the course of the seventeenth century—namely, an aesthetic of learned production, the active discussion of principles for ordering excerpts.[1] This chapter will trace the divergence and increasing disparity of the two indexical situations, leaving aside for now the library and its cataloging rooms in the late eighteenth century, in order to turn to the arrangement of the material in the solitary scholar's study. The chapter proposes a genealogy that ranges from liberal praise for assembling excerpts (J. J. Moser), to the poetic and poetological extension of the technique (J. P. F. Richter), and to its peculiar culminating in its characteristic silence (G. W. F. Hegel). First, the discussion of index cards should be articulated by dint of a functional and terminological difference—namely, the distinction drawn between (1) the gathering of material that is the basis for library catalogs and (2) the assembling of notes for texts to be written by a scholar's hand, which in turn will request admission into library catalogs.

The Scholar's Machine

No doubt the task of a library catalog consists in referring to *all* addresses of available books in as complete and logically consistent an order as is feasible at any given time.[2] Questions to the catalog—whether asked by the mediating librarian or later by readers themselves—customarily comply with this general schematized form: whether and where a text is found (author catalog), or which text can be found in the stacks (subject catalog). Thus, the catalog may be expected to be able to answer if the pattern is followed, regardless of how peculiar a query might be. In other words, the library catalog serves as a collective search engine (figure 4.1). Its data input comes from numerous sources, but it always works in accordance with strictly regulated instructions, so that it can be queried by anyone. In media theory terms: the communication structure of a collective search engine obeys a "network dialog."[3] A bibliophile notes in 1915:

> Attempts have been made to compare the catalogs of the large book repositories, the public libraries, to a herbarium of many volumes, where neatly ordered dead plants gather dust. Yet these directories of our large book towns tell their users much more than any herbarium could ever say to someone coming from a botanical garden. And if a lover of intellectual vegetation invests in a directory, it is bound to be a somewhat confessional piece; *the description of one's own garden always looks different* from the descriptions of the most agreeable things that grow in distant and foreign regions.[4]

The difference between the collective search engine and the learned box of paper slips lies in its contingency, and the resulting possibility that queries in one's own terms can be posed to the strange arrangement. While a search engine is designed to register *everything* randomly, the scholar's machine makes the determination whether or not to record a piece of information. This power of selection defines its idiosyncrasy. The materials accumulated over time by the herbal expert, the insider, can be allowed to form deviant systems that would be indecipherable to other systems, while nonetheless obeying the command that they always yield answers to the questions of their partner. The architecture of the idiosyncratic scholar's

Figure 4.1
A search engine. ["If you search the internet, you need to have the right tool."]
(From *Frankfurter Allgemeine Zeitung*, September 8, 1999, p. N3.)

machine[5] requires no mediation for, or access by, others. In dialog with the machine, an intimate communication is permitted. Only the close and confidential dialog results in the connections that lead an author to new texts. When queried by the uninitiated, the box of paper slips remains silent. It is literally a discreet/discrete machine.[6]

With the increasing separation of the scholar from his formerly common professional service *as a librarian* (as it occurs for example in Wolfenbüttel in the transition from Leibniz to Lessing and their respective activities), the author also necessarily withdraws from access to the library. His formerly direct control over masses of books wanes and becomes the domain and occupation of librarians. Scholars are forced to compile personal catalogs after their own fashion to organize notes and excerpts.

Material gathered and compiled contains a small library, diminished in complexity. It *sorts addresses* so as to *address thoughts*.

If, in the end, the catalog movement successfully overcomes the personal memory of a localized library and takes its place, then the requirement to serve as a commemorative place for books is also relinquished. The *catalog* functions not as a mnemonic device for writings lost in storage, but merely as a formal representational structure—namely, an interface. Assembled by professional librarians—who since Ebert are under orders to refrain from engaging in their own careers as writers—the library catalog operates only as a guide for visitors to the library.[7]

By contrast, the scholar's machine allows two different applications. First, it serves as a memory aid.[8] Bit by bit, it receives excerpted materials in order to fix them in a suitable place. Faithful to the adage that *stored* means *forgotten*, it offers an arrangement against the irreparable loss of the addresses that in turn point at content.[9] The second, more serious application moves the scholar's machine fully into the position of textual production. For it not only reliably reproduces everything the scholar gradually invested in it, recalling the extended present back to the time when each entry was made. Provided that the scholar knew how to tie new material together with the existing stock of excerpts, and marked connections and associations to similar texts and themes, the scholar's machine *as a text generator* delivers these very connections by branching out into forgotten memories as virtually new, served up as well as unexpected connections. The apparently insignificant, but regularly marked cross-reference yields rich profits when its recombinatory linkages enrich the power of the excerpts with chains of references.[10]

As a result of unilateral liberation from the bound book, freely interconnected slips of paper expand the intersections and so increase the connectivity of possible relationships. Thus, the material of the scholar's machine awaits skilled interrogation in the shape of addressable aggregations of paper slips kept in appropriate boxes. The first bureaucratic order of 1495 under Maximilian I already recommends the use of *partitioned* cabinets as storage places for files.[11] In Placcius's anonymous treatise of 1637 and Harsdörffer's *Erquickstunden* of 1653, these professional pieces of furniture appear miniaturized, adapted to serve as containers for paper collections. A refined blueprint for such an excerpt collection, proudly presented by a prolific writer of the eighteenth century, is discussed as an example in the following episode.

Genealogy: Johann Jacob Moser and Jean Paul

Gulliver saw a machine in Lagado that mixed paper slips to such an extent that anyone they were read to could not tell whether they were read from an ordinary book or not.
—Jean Paul, *Selections from the Devil's Papers*[12]

One of the best-known and most prolific lawyers of the eighteenth century, Johann Jacob Moser (1701–1785), mentions his box of excerpts as a method for collecting material for future writing:

I write all my findings on single little sheets, of which I make sixteen each from each sheet of paper, two from every octavo. Now if I find something someplace, and I believe that it might serve me in the future, but it is perhaps in a book that is not mine, so when I might need it again I would not have easy access to it, I copy it on a half-octavo, or extract from it; especially if it seems important. […] I had a carpenter furnish wooden boxes that each hold two rows of paper slips, divided in the middle by a thin piece of wood. The width of these boxes is such that it accommodates the width of a half-octavo slip; the height is equal to that of a half-octavo slip, and the depth or length is about one foot, so that each box can hold roughly a thousand slips.[13]

These stipulations will become the model for subsequent examples. Even around 1900, the standard box holds a thousand slips per cubbyhole, readily replacing the book as a format. In his relatively early challenge to the book, Moser limits himself to a not particularly well-balanced, yet compelling dialectic to persuade his readers of eight "advantages of this kind of collection," admitting only one disadvantage of the paper slip box. First, all topics remain together, but not written one after another as in a notebook—notebooks having sometimes too many and sometimes too few pages for their respective topics. Second, disorder can be avoided by insertion of paper slips, which books do not allow. Third, rearranging an entire text, adding new chapters, or removing old ones is easily accomplished with paper slips, while notebooks remain unchangeable. Fourth, the collection of paper slips requires little space: the user can extract paper slips, arrange them at will, and put them in a small box instead of being surrounded by numerous notebooks. Fifth, anything excerpted can be used in writing as it stands, instead of having to be copied again from a notebook. Sixth, the elaboration of a new piece of writing results more easily from slips of paper than from handling books. Seventh, the procedure is better for selection, or even concealment, of information: individual slips can be extracted and their information communicated (for example, to a copyist) without having to reveal a notebook's entire contents.[14] Finally, a flexible order offers the superior opportunity to add supplements in an orderly fashion. Only one "discomfort of this kind"—namely, the sum of all a librarian's fears—speaks against maintaining a collection of excerpts: "It is true that single slips of paper are more easily lost or scattered than notebooks: however, because I keep my paper slips in their own repositories, just like my books, and my people know that they may not lay hands on them, I have not suffered any adversity or accident with even a single paper slip."[15]

The basic value of the apparatus can hardly be overestimated.[16] Without this aid, Moser might have, by his own estimate, produced only half of his over 500 publications.[17] Looking back on his rather eventful life, Moser points once more to what he owes to the paper slip technique.[18] Indeed, his *Autobiography* contains an abbreviated version, appended to the second edition of 1777 as §46, of the listed advantages his method offers.[19] It is this section that regularly serves in literary history as reference for Jean Paul's organization of his excerpts.[20]

The year 1796 saw the publication of Jean Paul's *Quintus Fixlein*, with the full title *Life of Quintus Fixlein, Drawn from Fifteen Boxes of Paper Slips*. It is no coincidence that the story mentions as the "godfather and precursor of your slips of paper in boxes" a "Herr von Moser," both in the self-referentially titled *History of the Preface to the Second Edition* and in the second box of excerpts, which contains the memories of Fixlein's youth, stored on slips of paper. The title thus refers to Jean Paul's own mode of textual production as being based on Moser's methods.[21]

A narrative in Jean Paul's *Pocket Library* of the same year mentions an opposing technique not influenced by Moser: "Just excerpts. In the beginning I caught two, three oddities like butterflies from every book, and fixed them with ink in my scrapbook."[22] Thus, the media competition steps into the ring of contemporary controversy.[23] Dancing master Aubin, by profession a teacher of movement, advises his students to "at the same time read a lot and remember a lot," yet recommends, counter to the eightfold advantages of *thinking in boxes*, the use of notebooks.[24] This recommendation disturbs the consistent line of genealogy that directly links Jean Paul's poetological materiality to the creation of the physical arrangement of scholarly boxes. In addition, research on Jean Paul has delivered unequivocal evidence that Moser's procedures were in fact eclipsed by the use of notebooks.[25]

Nonetheless, poetic and scholarly production are reunited in the question of how to work with the assembled excerpts. Reviewing excerpts at regular intervals so as to remember not merely the citations, but the material as a whole, according to the Baroque rhetorical tradition—both Moser and Jean Paul declare this method their proper production aesthetic.[26] Moser describes the rereading of excerpts in detail to generate a new text by means of productive recombination:

Then I take all those slips of paper with the excerpted or otherwise used references, and put them between the divider and the octavo sheet that contains the rubric it belongs to, though not in any particular order: once I have them there, I take one chapter after another from the box, bring the slips of paper into a specific order, and add marginalia.

Then I go through my entire stock of books, and look in the volumes that seem promising on the subject, in their index or elsewhere, and note down other findings onto new paper slips of half an octavo sheet, before inserting them into the sorted slips.

Once I have come thus far, I am confident that what I can and should say is correct: now I go through it and reflect upon what is missing or could be improved, or where my own thoughts might or ought to be added, etc., which I then manage to do.

Also I must not forget to note that I never write more on one slip than can be found in this one text in one place, or in all my works in one sequence: but where a fresh line begins, I start a new slip, or a half octavo sheet; that way I will not run out of space if I have to add or correct, and I do not need to rewrite much.

Finally, I check the whole piece once again to see whether anything can be improved in writing and context, etc. Then I note down the marginalia of those paragraphs onto an octavo sheet containing the rubric of the chapter, under the term "content"; I number every paper slip, and then I hand the paper slips to the censor or the printer, without having to copy them again.[27]

Supplemented only with a table of contents and marginal notes, a final copy in the shape of a box of selected paper slips extracted from the collection goes to the typesetter—without having been rewritten. What comes back are the new book and the box of paper slips, to be returned to their respective places—namely, the excerpt cabinet and the bookshelf. Hence, the data circulate between two different agencies. The excerpt is selected from the book—to be made, as an excerpt of the excerpts, into yet another book. Jean Paul's dancing master distills this doubly selective funnel feature into a suitably liquid analogy: "The main thing is that I make excerpts from my excerpts, and distill the spirit once again. I may read them, for instance, only because of the article *on dancing*, or on flowers, re-moving this in two words into smaller notebooks or registers, and thus I fill bottles from the barrel's content."[28] It is no accident that this metaphor, which offers a poetic formula for the poetological procedure, borrows from a context of fluidity. Had there been no apparatus for the interconnection of liquifiable excerpts prior to Jean Paul, he would have had to invent one in order to bring the "core of his method"[29] to literary maturity. The idiosyncrasy of this style is based on a material arrangement that is able to create an incessant flow of comparisons. This working method relies entirely on the excerpt machine as text-generating prelude.

There is no lack of analogies for this idiosyncratic scholarly machine.[30] The leitmotif of the box of paper slips as an arrangement of books and their representations is the card or war game. As Jean Paul notes laconically in an afterthought to the catching of an excerpt-butterfly: "I drafted my recruits from all sciences."[31] On the other hand, Moser finishes his list of the advantages of his method as follows: "Indeed, a work compiled in this way looks a bit like chaos or a game of cards, since everything lies atop everything else, and if such a work were to remain unfinished, no one who does not understand this manner of working could finish it: but it appears to me more like a military regiment on a parade ground: before the parade begins, the soldiers run about regardless of order or beauty; yet when the signal is given, they take their guns; and every man is readily and most favorably positioned."[32]

Elsewhere

He excerpted continuously, and everything he read went from a book right past his head into another book.
—Georg Christoph Lichtenberg, *Sudelbücher*, notebook G 181, 1779

Because this study treats completion at most as a marginal objective[33]—I could never claim to represent the complete and exhaustive history of the index card—suffice it to indicate here the subsequent lines of development of the disposition outlined above, as the inevitable arrangement of prolific scholarly writing.

In 1785, when Hegel is fifteen years old and still in secondary school, he diligently begins to inscribe loose sheets of paper with excerpts.[34] In 1785, Johann Jacob Moser dies in Stuttgart, leaving behind, besides his boxes of index cards, his instructions on how to build a replica of his system.[35] Just as Jean Paul began his collection of excerpts at age fifteen in 1778, so it seems reasonable to count Hegel among the Swabian filiation of the readership of Moser's autobiography.[36]

Hegel maintains this proven production principle throughout his life, though making, as Friedrich Kittler observes, one decisive and far-reaching change. The reformulation of excerpts into new texts transforms the copyist into an artist. In Hegel's *Phenomenology of Spirit*, the absolute spirit appears as a "hidden box of index cards," because his excerpted foundation vanishes in the unstated erasure of references.[37]

In October 1822, Heinrich Heine visits Hegel; in October 1824, he visits Goethe. Two years later, Heine writes in the second part of his travel reports:

In all preceding chapters there is no line which does not belong there, my writing is condensed, I avoid everything superfluous, I sometimes even pass over the necessary: for instance, I have not yet quoted substantially—not minds, but authors—and yet, quoting old and new books is the main pleasure of a young author, and a few learned quotations adorn one's whole person. Do not believe, Madame, that I lack acquaintance with book titles. I know the tricks of influential minds who cherry-pick and choose quotations from notebooks; I am quite versed in the ways of the world. If need be, I could borrow quotations from my learned friends. My friend G. in Berlin is a veritable Rothschild of quotations and would be pleased to lend me some of his millions, and should he not have them in stock, he could easily borrow from some other cosmopolitan mind-banker—however, I need no loans, I am a man in good standing, I am able to consume 10,000 quotations annually, yes, I have even invented a way to pass incorrect quotations off as real ones. Should any wealthy scholar, e.g., Michael Beer, want to buy this secret, I would entertain a sale for 19,000 thalers; I take offers. There is another invention of mine I do not want to hide, and for the greater benefit of literature I will give it away for free: I think it advisable to quote all obscure authors by their street number. These "good people and bad musicians" (as the orchestra is addressed in the "Ponce de Leon"), these obscure authors always own one little copy of their own book, so that one needs only their street address to unearth the book.[38]

This description of traveling from prime addresses to others highlights the eloquently concealed counterpoint to the general discretion about the generation of texts in Goethe's times. Apart from Heine's clever coupling of book locations and residences, his flirtatious threat to reveal a silent arrangement imperils nothing less than the mystery of higher learning. "An open secret, under lock and key, / And only late revealed to the people," one reads in Goethe's *Faust*.[39] Yet by the same token, Heine's citation refers not by accident to the connection between quotations and means of payment, both kept in ways that suggest an analogy between a banknote and an index card. For the circulation of representations of money and thoughts obeys the same structural model of substitution, supported by notes to which in fact the same name is applicable.

Banknotes

Since the author speaks his works, the circulation of words can be connected to the circulation of money.
—Heinrich Bosse, *Autorschaft ist Werkherrschaft*

"Ten guilders of Viennese city banknotes which are accepted in all contributing community and cashier banks for cash, namely, for ten guilders," reads a banknote from the main cashier of the city of Vienna dated January 1, 1800.[40] These anonymous slips of paper are not yet money in the proper sense (legal tender), but merely a promise to respective owners to render cash upon submission of the slip of paper—the registered equivalent in gold or silver. This practice has its origin in England around the middle of the seventeenth century. It was not least the ongoing civil war that made wealthy people entrust goldsmiths and moneychangers with their capital for safekeeping. As vouchers, they receive so-called goldsmith's or banker's notes (see figure 4.2 for a modern version), which guarantee return upon demand.[41] As this system proves itself trustworthy, it gradually spreads, institutionalizing itself in banks that henceforth deal not so much in gold as in exchanges of paper. For the direct redemption of banknotes for cash allows the carrier (entitled to possession without producing any proof of ownership) to transfer the voucher to others. In this manner, the ephemeral

Figure 4.2
A banknote from 1873, by courtesy of Deutsche Bundesbank, Frankfurt am Main.

paper slip develops into an easy and comfortable means of payment—albeit limited to the geographical region of the bank that vouches for the exchange. From now on, it is no longer necessary to carry around heavy precious metals.

Using the comparatively valueless paper as a representation of value simplifies payments owing to easy transport and anonymous transfer. Moreover, banknotes offer the chance to add value. The promise to exchange paper for money is hardly ever going to be made good for all holders of paper at the same time. That is why a bank can hand out more slips of paper than it actually holds in precious metal. The capital thus gained can be deployed elsewhere. In contrast to legal deposits, freely circulating slips of paper earn no interest. Nevertheless, just like their relatives from the bibliographical apparatus of library catalogs, the slips of paper allow an increase of capital without being able to bear interest independently. Only their number determines the degree of possible profit: numerous slips of paper increase the monetary as well as intellectual yield gained by the institution called bank or library. A bank hoards capital, a library hoards books. Both are united in the desire to increase deposits and to enhance what is achieved in processing money and writing by means of circulating paper slips.

This is the decisive innovation for business conducted on the basis of paper slips: it is no longer the value itself that circulates (which is why the banknote does not count as money before the centralization of banks and the state monopoly on printing paper money), but only a substitute in the form of a written guarantee; what is exchanged is a plausible guarantee by a private bank that the holder will receive *the equivalent* of the paper on demand. This form of representing capital—by nothing but characters and figures that remain mobile in their materiality as paper units—resembles the structure that links bibliographical slips of paper with the books they refer to.[42] Ever since the extensive cataloging project in Vienna in 1780, it is the mobility of the card index that maintains access and thus the business of reading (and the subsequent processing of reading into new writing) under the conditions whereby huge numbers of books come into ever larger libraries. The historical coincidence of card catalogs with the gradual success of banknotes around 1800 as freely floating means of payment appears to be more than a mere historical contingency.[43]

However, to uphold the representation between a slip of paper and capital at every moment, one essential condition must be fulfilled: banknotes must be covered, the money backing them up must exist. The list of unfulfilled promises is long and includes several prominent examples: from John Law, whose first paper slip bank ran for only four years in France and was ruined by the war debts of Louis XIV in 1720; to the Austrian state bankruptcy of 1811, owing above all to a disproportion between the lack of precious metals and the abundance of available banknotes; and on to serious financial crises as a result of the Napoleonic wars in 1816 for the Bank of England, which had been economically stable since 1694.[44] Only Prussia remained unscathed by the new financial instrument: Frederick II hoarded the national treasury in barrels in a cellar, rather than handing it over as coverage for a paper bank.[45]

As a result of repeatedly failed attempts to lend banknotes the status of a stable currency and to strengthen the integrity of paper as money, banknotes were in bad repute. Alongside the honest efforts to set up paper banks as reliable institutions, numerous shady profiteers exploited the insufficient laws regulating private bank by spending more and more promissory notes without having secured their coverage, and in the short term greatly enriching themselves. Before any redemptions could be initiated, the bank was dissolved, and its operating authority as well as its capital disappeared without a trace.[46] Analogous doubts are deeply rooted in the librarian realm. Increasing admonitions point toward the fleeting nature of paper slips and their lack of reliability. In contrast to the permanence and duration of a bound catalog that guarantees finding a work among the abundance of material, books can easily fall into oblivion if the referring paper slip is lost or misfiled.[47] Without a guarantee of permanent access to banked capital, trust in paper slips is shaken. "In this respect our intellect is like a bank of issue which, if it is to be sound, must have ready money in the safe, in order to be able, on demand, to meet all the notes it has issued," as Schopenhauer remarks.[48]

Despite the skepticism that persists until the establishment of nation-states and state-guaranteed national currencies, payment with banknotes retains an essential advantage compared to payment with coins or gold: joint stockholders, as owners of the bank, profit from the promissory note system, and the depositors profit from interest earned, while the holders of banknotes must renounce interest. "The holders of the notes received no pecuniary yield for holding banknotes. Apparently they used only the fungibility of the paper slip. Yet only they possessed wealth with the highest possible degree of liquidity."[49]

Only this liquidity, the comfortable ability to let capital flow, empowers financial control within a system of economic exchange. Around 1850 in Prussia, demand for an increase in population mobility arises, giving rise to a new network for the circulation of human beings: a railroad system. The time had come "where the state must increase the available paper money."[50] Insofar as the "increase of the circulating medium,"[51] the reinforced application of paper substitutes for money, further liquefied the circulation of capital, one consequence was an overall acceleration of commerce. Again, it is not difficult to draw an analogy to the business of writing, the circulation of words: the mobility of card indexes precipitates the flow of texts. Library access to books is simplified and considerably increased—no time is wasted copying bound catalogs that limit the up-to-dateness of references. For readers in search of material for their subsequent activity as authors, this mode of liquification and fragmentation allows them to quote more easily—faithful to Montaigne's motto to "let others say what I can say less well"—to add quotations to their own collections and assemble them into new texts according to Moser's technique of weaving.[52] A sufficient amount, mindfully collected, arranged, and put into circulation, can then create the intellectual fluidity that makes Heine's friend G. in Berlin a Rothschild of quotations.

Even today, access to written sources is granted by thoughtfully cultivated databases. Yet other paths lead to the sources of wealth. For the current card bank of the sciences is not only a database of electronic references to books. Also on call are *citation indexes*, databases that indicate the frequency of references, that is, how often someone has been quoted and by whom.[53] This special mode of accumulating cultural capital delivers indicators and measures for another form of value added and hints at the current market value of an author, thus providing guidelines for future speaking fees, consulting honoraria, or tenure negotiations.[54]

The model for free-floating representation is "the circulating medium whose proper circulating in the financial body is no less important than human blood."[55] Monetary circulation and circulation of thought on paper find their analogy in William Harvey's discovery of blood circulation in 1628. "Money does not pulsate in the financial veins with the same regularity as in humans, but shares with the latter that a momentary blockage can cause danger or death."[56] A disturbance in the materiality of the flow of thought, one might add, is the threat of outdated, missing, or faulty index cards. It is the insufficient quantity of index cards that leads to effacement or malnourishment. "I feed the computers with my data." Only in their sheer abundance can citations create something new. The database emits, donates blood, gives life, on a massive scale. The circulation of index cards thrives on abundance as well—the more the better. "I am the database. Bleeding in the crowd." [57]

Balance Sheet

In their time, men like Moser could proudly refer to their index cards as a text-generating technology, contributing to the Enlightenment with an almost uncanny production rate. Yet around 1800, with the blossoming of the European idea of genius, this light dims, and production aesthetics undergo a fundamental change. From now on, painstakingly produced drafts go unmentioned, veiling the writing process in the darkness of a productive sleep.[58] Darkness keenly protects the trade secret of textual genius. Mention of index cards as a production aid is generally banned: external tools are not supposed to be required for the writing of great works. The *genius tempi* denies its *locis communes*. Heine's tempting offers to divulge knowledge about commercial networks of plentiful knowledge and, beyond that, to denounce the secrets of scholarship, including successful forgery, are directed precisely against this glorification of genius.[59]

While the point of the scholar's machine around 1800 is to make its own discursivity disappear, the exact reversal of this reticence occurs with the gradual establishment of the card index. Despite harsh criticism,[60] book collections give cause for pride, reemphasizing the technical basis of literary production. The effective enforcement of modern storage systems in the course of the nineteenth century also allows the index card to celebrate its return as a literary and scholarly text generator, possibly enhancing the life of many a mediocre author.[61]

In Praise of the Cross-Reference

The basis of everyday indexing is copying or excerpting. As such an excerpt offers only one small segment of the source text at any given time, it refers to a larger context that remains unincorporated, though this context is at least adopted as an address. In other words, the excerpt is a pointer that also refers to something else—and presumably not just to what is essential. However, excerpting alone does not make a collection of a specific productive strength. What is the use of the most painstaking copy if it cannot be brought into productive relation with other entries? What is the use of several pages of excerpts if they do not enter into a network of connections? In isolation, every index card is in peril of becoming a data corpse; to live, it must enter into relations with the remaining content. Thus, what single slips of paper require is cross-referencing. For only by means of cross-references do the disparate single entries, fed in at different times, constitute a web of evident relations, whether consciously drawn by the user or unconsciously put together by the machine.

Only through this skill does the index card box grow from a mere filing instrument to an author's assistant, or even—as we will see—into a regular communication partner during textual production. For the apparatus returns infinitely more than the user feeds into it. As soon as one regularly cross-references new input with older material, the index database blazes associative trails that may serve as clarifying creative prompts for different connections and unexpected arguments.

Now one might ask whether the connection logic (which merely plans to tie together, for instance, the same keywords with each other) is not too unrefined, perhaps producing inconsistent connections. Yet the box of index cards can counter even this objection and turn it into a productive response: Denis Diderot, for instance, intentionally planted incorrect cross-references in the *Encyclopedia*—in order to stimulate the mind.[62] Cross-references are the crucial feature of a network—even if they produce connections that appear to lead nowhere—for only one link just might offer, in another connection, at another time, with shifted research priorities, interesting new constellations.

Another feature is inherent in this developing network of successively expanding information: the aspect of surprise. On the basis of keywords and short forms, every point of the index card box can refer precisely to another. In contrast to a book with its fixed connections and unchangeable format defaults, every slip of paper represents a finite, extendable info-unit, an expandable, elementary piece of information that can easily be cited—for every index card carries an unequivocal address thanks to its position in the set order or in the form of a number others can refer to. "Every note is only one element that receives its quality only from the network of the references and back-references within the system."[63] With the aid of these connections, the user succeeds in tracking down new connections following the reference structure of entries, uncovering unintended readings. "The box of index cards yields combinatory possibilities that can never have been planned, anticipated, or conceived that way."[64] Consequently, surprises pop up, thanks to an unexpected reference to aspects not previously considered. So how does one furnish a box of index cards with the ability to surprise? Only over time, time being the feature that enables complicated structures to develop. These will emerge under the condition that the user consistently feeds the system with information—namely, in the shape of text modules, facts, fragments of thought, longer excerpts, even complete arguments, always binding them into the existing reference structure.

Index cards owe their potential for surprise to the reading effect. If the accumulated notes remind users of what they were thinking, and if the texts also exhibit associations and connections to the complex rest of the content, then the notes serve not only as a memory aid, but also as a comparative horizon, shifted by time. What surprises is not only the difference in understanding between this time and that time—ah, so that's how I took this at that time—but also the potential path that was not developed the first time the author noticed it, but looks promising when accessed again. The system of notes develops surreptitiously, as it were: a box of index cards yields not only the preparation for a new text, but also a mold for text yet to be written. The cross-reference creates, almost autonomously, a kind of argumentative surplus that is the true value added of a box of index cards, while incessantly helping the reader fix his or her impressions and associations. This connecting principle (celebrated by the apologists of the digital age as the so-called hypertext) was anachronistically employed by Niklas Luhmann in the shape of actual index cards in wooden boxes. While others had long ago resorted to feeding electronic databases, Luhmann, at the end of the twentieth century, still adhered to wood, paper, and ink.[65]

On the Gradual Manufacturing of Thoughts in Storage

As soon a box of index cards reaches a critical mass of entries and cross-references, it offers the basis for a special form of communication, a proper poetological procedure of knowledge production that leads users to unexpected results. The premise of this claim (and the basic principle of working with index cards since their "invention" by Konrad Gessner) consists in the fact that innovation never happens ex nihilo (be it a thought, or a text with delicate lines of argumentation, or the creation of an artifact), but on the contrary always includes a recombination of disparate or similar elements. In short, the production of innovations is always based on the fortified recombination of the existing.[66]

Twentieth-century practitioners of of index card techniques like Niklas Luhmann assume that the wood-and-paper apparatus can in fact be an equal and stimulating partner in communication; in fact, though, this idea dates back to a constellation described by Heinrich von Kleist as the "midwifery of thought" in his fascinating 1805 analysis *On the Gradual Production of Thoughts Whilst Speaking*. "If there is something you wish to know and by meditation you cannot find it, my advice to you, my ingenious old friend, is: speak about it with the first acquaintance you encounter."[67] The beneficial tension induced by the interlocutor's expectations immediately spurs new thoughts; the idea develops during speech. The mere presence of an interlocutor is sufficient; he or she does not need to do anything, nor offer witty responses as additional stimulation. "It is a strangely inspiring thing to have a human face before us as we speak; and often a look announcing that a half-expressed thought is already grasped gives us its other half's expression."[68]

Kleist's idea is that communication partners need a catalyst to attain clarity about the knowledge to be expressed. What does the mere presence of an interlocutor achieve? "The midwifery of thought"[69]—a term Kleist borrowed from Kant. Without a counterpart, intellectual bankruptcy looms, as in the case of Heine's friend G., but with a partner, wealth beckons. It is no coincidence that early collections of analects and excerpts from all readings date from the same era as the paper banks of the sixteenth century.

According to Kleist, the human face serves as a sufficient source for inspiration; a look that conveys comprehension of a half-expressed thought is sufficient. One could assume that looking at wooden drawers is hardly inspiring. However, substitute the "interface" between person and apparatus for the "human face"—and substitute for the simple word "gaze" the word "brush." For it is the light touch of index cards, the interplay with the silent interlocutor, that gives birth to a thought, prompting the interlocutor to actually speak and provide "witty responses" after all, to use Kleist's phrase. The box of index cards offers an interface that is more than just a stimulating sight, as the apparatus, upon the lightest touch, delivers keywords that stimulate the protagonist to further production of thought. Thus, a silent counterpart can grow into an actual interlocutor. The fact that the keywords offered seem by no means arbitrary is guaranteed by a widespread net of cross-references, for both partners. Bit by bit, during the course of their interaction, connections accumulate in the apparatus as a "kind of second memory."[70] And this second memory gains a certain amount of independence if it intervenes in the thought process of its thoughtful user more thoroughly than Kleist suggested.

Yet compared with a mere silent human face, the box of index cards offers another advantage as an interface. Its advantage consists not only in its ability to deliver precise answers to specific inquiries, but above all in its infallible ability to remember associations, to say nothing of the value added that is offered by automatic associative linking. Every input is preserved and retrievable, either as an isolated piece of information or as a building block for a larger line of argument. Getting involved in constant communication with such a second memory not only means trusting in the fact that the apparatus faithfully returns the stored information—there is also the reliable fact that information successively fed in over time will generate future knowledge. To use Kleist's words once again, "For it is not we who know things but preeminently a certain condition of ours which knows."[71] The apparatus retains these possible states. In its connectivity of preformed elements, it achieves a configuration of potential states of knowledge that are merely actualized by the user at a given time in certain combinations—when they are called up.

"The text knows more than its author," as one of the basic assumptions of philology has it. One could transfer this statement easily to the relation between boxes of index cards and their users. Text fragments held at the ready by the apparatus in their potential connectivity offer incomparably more connection points than the user is aware of at any given moment. Thus, the interface offers a range of possible connections, and along with them it delivers potentially new lines of argumentation. Storing states of knowledge and (via their contacts with the interface) helping to catalyze future thoughts, index cards know more than their author.

To what extent must this small apparatus history about the midwifery of thought remain theoretical, and to what extent can it help to illuminate the peculiar situation of a scholar and his private card index? One might say that the communication between database and user is purely theoretical, in the etymological sense of the word *theoria*, "view." For the boxes of paper slips allow their user an instant view, an overview of possible constellations or different arguments. The variety of the slips of paper opens a perspective onto different possible considerations at the same time, allowing one to see various mental constellations in their contingency. Theory is nothing else, at least in etymological terms. It is up to the users to commit themselves to a view, to select lines of argument or readings according to scientific practice as the basis of their own textual production.

At this point, we will leave the evolution of the scholar's machine behind, with its recombinatory creativity owing to the material nature of paper slips and wooden boxes—save only to note how this line of development leads directly to Arno Schmidt or Vladimir Nabokov as an apotheosis of literary word-processing with index cards and boxes of paper slips.[72] Instead, let us try to retrace the final transference of the library technology called *card index* into an economic field, where it will prosper in the history of technology. This occurs in America—on the one hand as an adaptation and transatlantic translation of index card technique à la Gessner in the newly established large libraries on the American East Coast, and on the other hand as a method with its own unique American genealogy. Untouched by the achievements of old European library technology, it is independently invented at Harvard in 1817, born from the spirit of sloth.

5 American Arrival

Have you a catalogue?
—William Shakespeare, *Coriolanus*

How, then, does the library card index reach the New World, and how does it also develop into a card index system for business use? On the one hand, American librarians travel around Europe throughout the first half of the nineteenth century to study—and later to import—library technologies for their own, rapidly developing libraries, particularly in New England. Well-known and influential American librarians like Charles Coffin Jewett (1816–1868), who maintains a close connection with Anthony Panizzi at the British Museum starting in 1845, or Joseph Green Cogswell (1786–1871), who studies at Göttingen for several years and later manages the famous library of Harvard College, exemplify the contemporary reception and dissemination of library science. However, this technology transfer between Old and New World relinquishes its European roots all too soon, and the paper slip system is soon claimed as a home-grown method. Yet this repression was unnecessary, for North American library history can lay claim to its own developments in matters of efficient cataloging. What follows is a reconstruction (from the archives at Harvard) of the independent "invention" of an old European paper slip technique in the New World. An eccentric protagonist stands at the ready; here we encounter a curious chapter in index card history that the standard American library history, concerned with its reputation, tends to keep under wraps.

Do Not Disturb—William Croswell

I have dearly paid for all the honor that success can give.
—William Croswell

On August 4, 1812, the Harvard Corporation hires a new library employee. His task for "not less than three months"[1] consists of cataloging the extensive book collection at Harvard College. "When my employment at Cambridge commenced, I professed no skill in the arrangement of Library Catalogues. I had other qualifications, and I expected that a proper Plan would be furnished. A Latin Plan had been compiled for the Library Catalogue by a former Librarian. This Plan was delivered to me; and about one fourth of the entire books in the Library were designated according to it."[2] As a former mathematics teacher who dabbled in astronomy and project management, William Croswell embarks on his career as a library assistant without the slightest experience with bibliographies or cataloging.[3] In his own words, his qualifications consist in the fact that he is "acquainted with ancient and modern languages, [trained] to write a fairly hand [i.e., a fair hand], and ready in detecting literal errors."[4] The work on the catalog starts well, but after only a few dull months of copying titles into a register, the execution of the simplest working algorithm falters. However, the gradual waning of bibliographical productivity is accompanied by a noteworthy process, and one that—albeit joyless for Croswell and his supervisor, Harvard College president John Thornton Kirkland—will result in a major achievement for American information management technology. William Croswell endows the largest library of his country and his time with a paper slip catalog that, as a prototype of the card index, will find its way into the offices and management systems of the prospering economy around 1900. "I examined the Library attentively. I opened every volume of every Set and noted whatever appeared irregular. I spent about three years in entering the entire books and also the Tracts in the four first Alcoves: the remaining Tracts were entered at Lodgings."[5] The book collection in Harvard Square in the heart of Cambridge, Massachusetts, is not only one of the oldest in the United States (founded 1639), but also the largest; around 1800, the library includes nearly 20,000 volumes.[6] Considering this fact, it makes sense that indexing the entire holdings of the collection within three months, and without the help of assistants, is virtually impossible. Which technique of cataloging does the completely inexperienced librarian Croswell select? First, he starts a library diary in which he enters each individual working step. This diary is extant in two versions.[7] One is more detailed, with an appendix of additional remarks, and was written as Croswell worked. The other version summarizes the events and was presumably written as a later justification. Both are marked by sporadic and short entries, particularly in comparison to the accurate and long-winded discourse of his successors, such as John Sibley's library journal of 1856. Nonetheless, they document Croswell's activities well enough.

In 1812, the library employs only John Lovejoy Abbot as its official librarian. However, Abbot leaves this position in 1813; the new librarian Andrew Norton soon assumes a more promising theology professorship, yet retains nominal responsibility for the book collection. As a result of his other assignments, the librarian Norton neglects to supervise his assistant Croswell, who can thus neglect his own work. In contrast to his initial diligence in copying book titles onto manuscript pages,[8] Croswell's zeal (and consequently his productivity) dwindle with the lack of direct supervision. The diary leaves no doubts: entries become rare and terse, as for example in this note from February 23, 1813: "M[onday]—23. At the Lib—do nothing." In the same year, December boasts one single entry, which adequately summarizes the prevailing activity of the unattended assistant: "Christmas." Over the following years, entire months are condensed into just one laconic term: "Nihil" (as in March 1817; see figure 5.1). Although this expression could be translated as "no special incidents," the other meaning—"did nothing"—remains a great deal more likely.

Meanwhile, Croswell does not remain entirely idle. Apart from occasional entries about swimming in the nearby Charles River, even in winter, and meticulously noted dates when the library remained unheated and thus too cold for any work, the library diary records isolated, intentionally vague bibliographical activities. However, in 1816, four undisturbed years after the beginning of his employment, Croswell is interrupted in his hitherto unnoticed labor by two unexpected incidents. After four years, the library still lacks a useful catalog. As President Kirkland's patience wanes, the pressure on Croswell and his lackadaisical cataloging attempts becomes more immediate. "The President told me April 20 1816 that it was expected that I should prepare for the Press."[9] Nevertheless, the request to hand over his catalog submissions to the printer passes by without a renewed deadline or serious consequences. That same summer, Kirkland announces a plan to rearrange the library according to the classification pattern of Jacques Charles Brunet. Croswell sees no recognition for his work—this time despite serious efforts that nonetheless do not seem to do

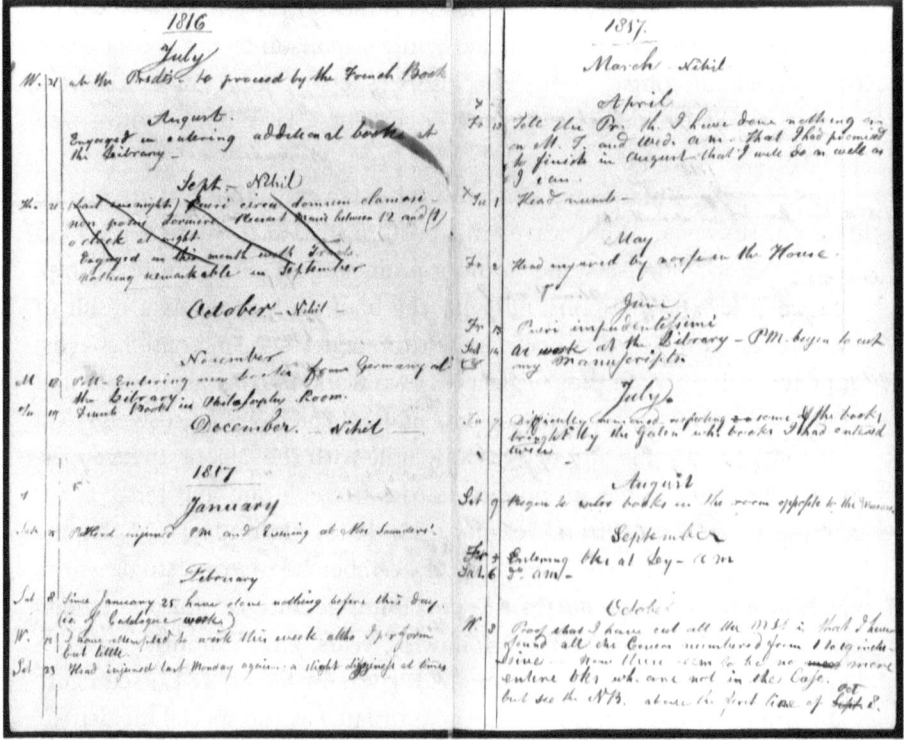

Figure 5.1
A double page from William Croswell's diary from July 1816 to October 1817. (From Harvard University Archives, Cambridge, Mass.)

justice to the assignment. "I designated the entire books throughout the Library by numbers, according to Brunet's method, and then it was rejected. This work of mine is contained in a folio manuscript of 180 pages, and may at any time be compared with Brunet."[10] Additional problems arise: the library's peace is disturbed once again, this time from a rather unexpected quarter. Presumably to justify his rather modest advances with cataloging, William Croswell informs the president about an incident that in a letter dated March 10, 1817, two months after the event, grows into an accident.

Rev. Sir.

I ought to communicate whatever affects the Catalogue business. On the 25th of January my head was extremely injured by the jarring of the doors. Since then very little has been accomplished: not a week's work in the whole.

If a suitable place can be procured I shall still hope for a satisfactory conclusion of the business.

With due respect

W. Croswell[11]

Having thus drawn a line of defense, as it were, in the form of sound insulation, Croswell follows up with numerous further complaints, protesting the traumatic effects of considerable acoustic disturbances.[12] Thus, it is hardly surprising that subsequent diary entries list more "Nihil" and "head numbs" than actual library work. Apart from a timid attempt in February 1817 to carry out his library duties—doomed to failure from the start—the diary reads as an anamnesis of the case of Croswell the headache expert, to borrow Kafka's phrase. The records remain a sick bay file up to one noteworthy Saturday afternoon, June 14, 1817. While the entry of the previous day, Friday the 13th, denounced a neighboring school class as the most recent source of noise and cause of disturbance ("Pueri impudentissimi"), the next day, emphasized by a special symbol in the margin, ☞, marks a sharp turning point in his career in particular and in American cataloging in general. For the first time since the creak of the library doors that triggered six months of disability, the diary announces: "at work at the Library." And after a dash: "PM. begin to cut my Manuscripts."

This fragmenting of manuscript pages commences the first phase in the creation of an American card index. From the predicament of having achieved little within the past three years (as a matter of sloth) and nothing during the last months (as a matter of illness, real or imaginary), William Croswell takes the bull by its horns. He recognizes the need to accelerate his work and tries to find the easiest method that will lead quickly and efficiently to results. What would make more sense than to bring existing material into a new form by cutting it up and rearranging it? As early as October 1817, the latest catalog appears to be finished. On Wednesday, October 8, Croswell's diary notes, "Proof that I have cut all the MSS [manuscripts] is that I have found all the Covers numbered from 1 to 19 inclusive—now there seem to be no more entire bks wh. [books which] are not in the Case." Everything that had been part of Croswell's manuscript is divided carefully into single slips of paper. These are arranged in alphabetical order, "disposed in a Case (furnished by the Librarian)."[13] In one of the last extant letters, a draft rather than a letter and presumably never sent, William Croswell describes the construction and order of his card catalog once more in greater detail: "I noted the bks on my Mss by numbers acg [according to] this Plan. And then cut my new Mss into Slips each Slip relating duly to one book. I cld trust none but myself to cut Mss. I marked the Divisions in the Case orderly by Numbers; and proceeded to place each Slip in its proper cell."[14] Conscientiously ordering the slips of paper into slots in the box forms the second phase in the development of this homegrown card catalog. As in Vienna in 1780, this second stage abandons the temporary status of index card arrangements and puts them permanently to work. In the absence of alternatives, Croswell's card catalog, born from a lack of time, had to serve as the regular Harvard College library catalog for at least three years.

Although Croswell begins to "form pts [parts] of the new Cal. [Catalog] by pasting briefs of the s.Kd [same kind] in separate Mss. regarding alphabet order, and also the order of time," a large portion of the index slips still remains in the box. Meanwhile, President Kirkland prescribes a different plan, confronting the library assistant with considerable new problems regarding the conception of a catalog, demanding the rearrangement of the entire system. Croswell suspects nefarious intentions and harassment on the part of the president, and he develops a strategy. "There is much to be commented in the Plan that was given me and some tendency to perplex and mislead. I formed a new Plan without needless deviations from the other."[15] A later letter to the college administration refers once more to the process that is expected to explain his delays: "After I had lost much time and labour, I concluded it unsafe to trust this Plan implicitly in any part. I formed a fourth Plan. This was a work of time. I succeeded at last in disposing all the remaining Slips in my manuscripts under apparently proper heads, excepting a score or two, reserved for farther examination."[16] In view of the increasing pressure on Croswell, it is only due to the flexible order of index cards in the box that Croswell is able to produce a catalog. However, the catalog itself, the fruit of this technique, is held in low esteem by his superiors. In 1824, Croswell sums up his mixture of pride and bitterness: "It was *now easy* to complete the Catalogue according to *any* approved Plan. The Slips had been regularly arranged, and indexes prefixed to the Manuscripts that I had finished. The whole had been written with great care, that no person might mistake a single letter. Such was my mode of paying the debt of gratitude for the employment given me."[17] Although his results are denied any thanks or acknowledgment by his employer, the story of William Croswell and his painful labor that led to the reinvention of an old European library technique does not end here. After Croswell's many years spent at the library, his catalog actually does come into existence; nevertheless, its perfection and completion turns out to be just as laborious, and should be outlined here briefly.

From autumn 1819 far into the next year, Croswell's productivity reaches a new low. Apart from a few exceptions, the diary registers the brief and unequivocal "Nihil" month after month. Croswell remains impassive, though by the middle of 1820 Kirkland's patience seems to be exhausted. In August, Kirkland authorizes a payment of $30 to Croswell, but under one condition: that by the end of the year he present the agreed-upon catalog of the entire library's holdings. Only six months later, Croswell turns to Kirkland once again, eighteen days before the deadline:

Rev Sir,

I should be glad to receive an Order for thirty dollars. No slips of value will remain in the Case after the present month excepting a score or two that are attended with some uncertainty. There must yet be adjustment and correction.

The Catalogue of 1790 opened the road.[18] I proceed with moderation because I dread a return of giddiness.

Your obedient servant

W. Croswell[19]

Meanwhile, time pressure makes the weary cataloger desperate. Croswell, evidently unaware of the polyhistorical indexing method, applies to the impatiently awaited catalog precisely the same method that Konrad Gessner had described and presented 269 years earlier.[20] Making sure not to forget a single paper slip, Croswell glues every scrap of paper on empty, bound sheets of paper. He uses index cards copied by himself from books on the shelves, as well as slips he cut out of the printed 1790 catalog. The result is a hybrid card catalog, resembling the designs of Gessner and Placcius, and consisting of printed and handwritten slips of paper (see figure 5.2).

The initial assignment to write a new catalog for the Harvard College Library seems fulfilled. Though it is presented after no less than nine years, including "about forty folio volumes on more than 700 pages," and just in time for the very last deadline, William Croswell holds back five of the forty volumes for safekeeping in his apartment.[21] "I cannot yield to others the credit of such severe labour. The manuscripts must remain in my

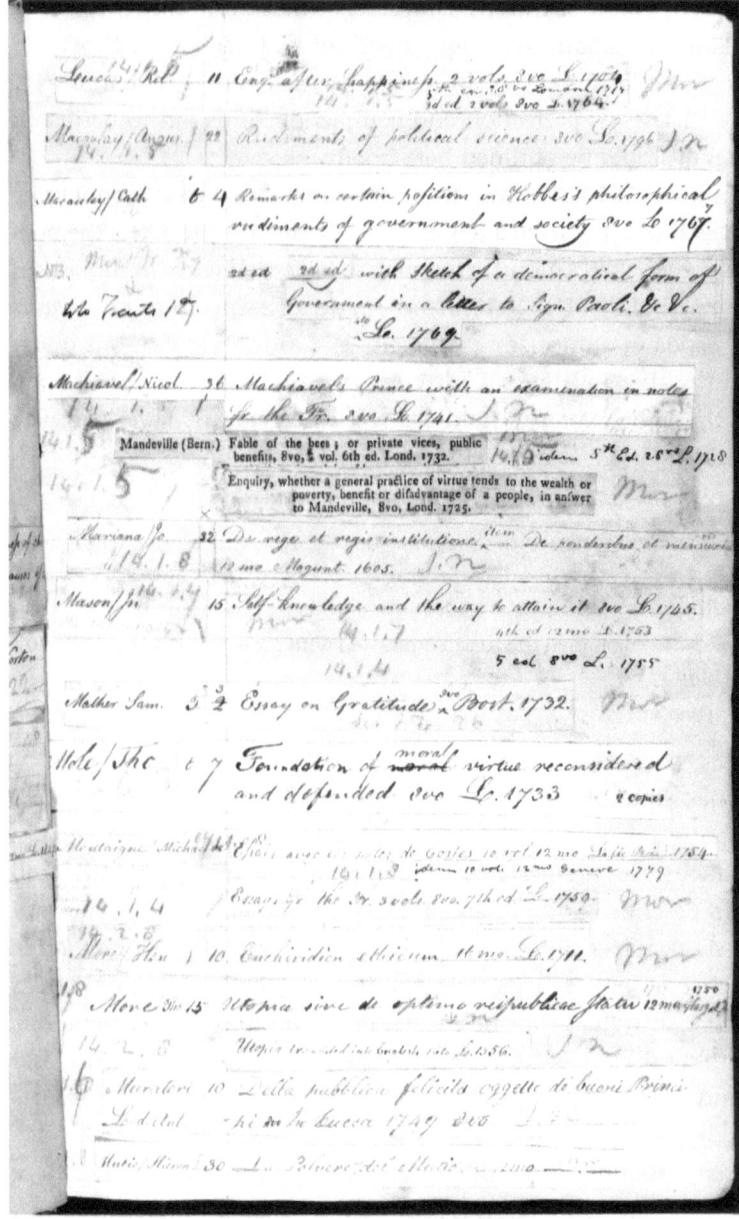

Figure 5.2
A typical page from William Croswell's catalog, containing printed as well as handwritten slips of paper. (From Harvard University Archives, Cambridge, Mass.)

hands. I shall consider them a sacred deposit."[22] Early in 1821, Kirkland explains that Croswell can expect no further remuneration; he is dismissed from the Harvard College Library. Yet Croswell still holds the Harvard Corporation liable and thus keeps the remaining five catalog volumes as a "sacred deposit" for additional payments by his former employers. "I had taken more pains with some of them than Authors always take with their published works of the same extent."[23] After some costly petitions[24] to judges unsuccessfully seeking an injunction, Croswell yields. The papers are confiscated in March 1821 "by a legal procedure."[25] As a result, the library diary that listed Croswell's activities diligently ends with the farsighted remark "Th. April, 5—Last day (credo)."

But there is a sequel. In spite of his dismissal, Croswell demands a pension from Harvard amounting to $762.15, according to his own calculations. For after his failed attempts to earn a living as a project manager, Croswell suffers from acute poverty. He develops numerous concepts for books, such as an edition of the collected works of Horace in a print-on-demand scheme, yet he cannot find enough subscribers in advance. Similarly for the account of his by then extensive bibliographical experiences: *A Collection of the Most Approved Systems of Bibliography* and *General Plan for Classing Books, with Remarks and a New Plan* are never printed for lack of interest.[26] Nevertheless, up to his death in 1834, William Croswell continues to send his reparation and pension claims to Harvard at regular intervals. For over thirteen years, he writes the same letter time and again with little variation, in the end with a shaky hand, recounting the history of his self-sacrificing contributions to the Harvard College Library, and not once receiving an answer.

The long lapses in Croswell's activity, noted with "Nihil" in his diary, form the conditions for a reinvention of the card index structure under time-critical conditions and floods of books. Without his vehement resistance against continuous work in the library over a period of almost nine years, William Croswell would never have been in a situation where he had to prepare a catalog of more than 20,000 books in less than six months single-handedly. In this predicament, in search of lost time and misplaced slips of paper, he invents a system that affords mobility, flexibility, and reordering of the respective units. His achievement for early American library and indexing practice is more than merely a short-lived temporary indexing tool. Croswell's catalog arrangement, independent of (and yet close to) Konrad Gessner's ideas and similar to the first Viennese card index, suggests a practically permanent application of the paper slip collection in specially crafted boxes. Presumably one may credit President Kirkland and the Harvard College Library with making sure there would be a catalog, albeit one in a rigid order with slips of paper glued onto empty, bound pages.[27] Only their insistence temporarily prevented the introduction of a permanent card index. However, the foundation of a modern cataloging system was laid, and it did not go unnoticed.

Early Fruits and Dissemination

William Croswell's direct successors adapt his system and apply his paper slips to their own enterprises. Three stages are worth mentioning, before Ezra Abbot in 1861 constructs his influential variation on the modern, publicly accessible card index. Each of these three phases is closely related to the succession of library directors.

After the hapless Croswell's dismissal, Joseph Green Cogswell is appointed full-time librarian, and is expected as a matter of course to come up with a new and improved catalog. Cogswell had just returned from a long stay in Germany, where he not only met Goethe, the famous supervisor of the Weimar and Jena libraries, but also became familiar with the most advanced achievements of European library science, in particular at the famous university in Göttingen.[28] For several years, Cogswell was instructed there by Georg Behnecke,[29] a student of Christian Gottlob Heyne, who himself was one of the leading librarians in eighteenth-century Germany.[30] For Heyne, the era of the catalog as a main library instrument was yet to come. For the time being, he concentrated his efforts on improving everyday library transactions, focusing on the arrangement of books on the shelf and corresponding classification systems within the installation. Hence, it is barely surprising that Cogswell's first official act consists of a complete rearrangement of Harvard's books. Only later does he turn his attention to cataloging questions, adapting the bound-book catalog that he came across on his visits to German libraries. Meanwhile, the basic card index principle is on standby in the Harvard College Library. Cogswell discovers Croswell's catalog (and probably also his boxes); moreover, he recognizes the potential of an order that stores bibliographical data *permanently* on slips of paper. There is no doubt that he makes extensive use of this legacy for his own cataloging work. "Seventeen boxes of folio half-sheets, each containing the name of a single author and abbreviated titles of his works" serve once more as a card index and produce the considerable advantage "that the work is done for ever; it may be increased so as to embrace all the books ever printed, without requiring any part of it now done, to be done again."[31] Although occasionally half a folio page contains more than one entry or summarizes several works of an author at a time—strictly speaking a step backward, as an efficient catalog would allow for one sheet of paper per text—this catalog has one of the indispensable features of the modern card index: the paper is of uniform size. In addition, each entry contains every piece of information needed to describe the represented text, so that a new catalog could be compiled simply by reordering the slips of paper without having to consult the books on the shelves. "Over all, as a master record, was the sheet catalogue, an alphabetical record of everything in the Library. This, too, was based on Croswell's work."[32]

After only two years, in spring 1823, Joseph Green Cogswell quits his job at Harvard; his work on the general catalog is still incomplete owing to the library's low budget.[33] Cogswell does not abandon his profession for too long, though. In 1842, he takes charge of the Astor Library in New York, precursor to the famous New York Public Library. Meanwhile, Charles Folsom assumes the position of librarian at Harvard. He familiarizes himself with the cataloging that has served the library in prior decades, but mistrusts the unconventional card-indexing procedures. During his three-year term in office, Folsom meticulously transfers the "sheet catalogue" back into a bound catalog.[34] But in doing so, he apparently learns a lesson: around 1850, Folsom personally brings the idea of cataloging on index cards only to the attention of Charles Coffin Jewett, who is faced with the task of cataloging the extensive collections of the Smithsonian Institution in Washington, D.C.

The Harvard "sheet catalogue" falls into oblivion, until several years later a further attempt is made to put together a new printed catalog. In 1830, this catalog is printed and published on the basis of the very same "sheet catalogue," serving as master record. In the following year, Thaddeus William Harris (1795–1856) takes over, and despite the new bound catalog, he occasionally consults the loose sheets of paper. Harris seems convinced that this specific form of processing has considerable advantages: from this point on, new entries are again recorded on slips of paper. In addition, in the annual report on the state of the collection in 1840, he recommends, for every future catalog of the Harvard College Library, once and for all: "That the Corporation should authorize a slip catalogue to be made, consisting of the title of every work in the library, on pieces of card 6½ inches long and 1½ inches wide; such catalogue being much wanted when books are arranged for the Annual Examinations to indicate missing books, and would also be extremely useful in facilitating the re-arrangement of books in the new library, and would serve for various other useful purposes hereafter."[35] After Charles Folsom's reign in the library, a regression in terms of cataloging, progress is imminent in the shape of the modern card catalog: after standardizing the cards, Thaddeus Harris transfers them—just like Abbé Rozier had done—onto a more stable foundation: instead of employing loose sheets of paper, he uses cards with greater durability to strengthen the catalog against wear and tear. "The only step remaining was to paste these slips on cards of uniform size, instead of sheets, and the result was a card catalogue."[36] For easy access, cards are kept in wooden drawers. This catalog lasts through 1912, in active use as master record for nearly seventy years.

A final step in the development of the modern card index, unrestricted access to the catalog for a common readership, finally occurs in 1861 when Ezra Abbot (1819–1884), a highly regarded Bible scholar and library assistant at Harvard under John Langdon Sibley, establishes new features. After more than twenty years of practical application, the index cards at Harvard have proven feasible for internal library use. Yet Abbot recognizes the necessity that individual readers attain access to the collection as well. Having gathered library experience in high school (as well as during his work on the detailed bibliography to *Doctrine of a Future Life*), Abbot is unhappy with the fact that new acquisition information circulates only internally and is rendered inaccessible to the general reader.[37] According to his didactic axiom of independent study, Abbot orders the internal catalog to be opened to an interested readership in the near future. To this end, Abbot presents a new plan: "I call attention to this point, because, as will hereafter be seen, in the new plan which I have the honor to submit of a catalogue designed to be accessible at pleasure to all who use the Library, this delay will be avoided; and the results of the labor of each week, or even of each day, in cataloguing, may be made immediately available."[38]

The master record—namely, Harris's 1833 catalog, accessible until then only to librarians—once again serves as a basis for the card-by-card generation of two new "indexes," an alphabetical one and one arranged by subject. Acquisitions are registered first in the master record, then on cards for the other two registers—including cross-references. "[The] work which is once correctly done on these cards is done for ever. [T]he cards that have been written, representing the new and important additions, may be inserted in their proper places in the drawer."[39] As an additional aid to the reader, the new catalogs receive a special ergonomic design that will considerably determine the shape of the first card index systems a few years later:

> The cards composing this catalogue are kept in drawers, twenty-eight of which occupy the upper part of a case, and are arranged in seven tiers, being placed at such an altitude that the highest drawer is not too high nor the lowest too low to admit of a convenient examination of its contents. Each drawer is about 15¼ inches long, 10⅝ inches wide, and 2¼ inches deep, inside measure, and being divided by a thin partition running lengthwise through the middle, contains two rows of cards. It is prevented from being pulled out accidentally by a wooden button screwed on the inside of the back of each half-drawer, and, when turned up, projecting a little above it. The drawer, on being pulled out, is therefore stopped by the buttons when they reach the horizontal partition in front on which the drawer above it rests; if the buttons are turned down, the drawer may be taken out.[40]

Moreover, each drawer bears a label, indicating its position in the catalog. "The cases of which these drawers form the upper part are each about 4 feet 3 inches long and 19½ inches wide, and stand on casters. [...] They are closed at the back, and the space in front below the drawers is left open, to be occupied with books, so that no room is lost."[41] Inside the drawers, wooden blocks are inserted to support the cards at an angle, enhancing the legibility of cards, titles, and details. The dimensions of the cards are thoughtfully conceived and strictly standardized: "I would propose to have the titles written on cards, about 5 inches long and 2 wide, of such thickness that they can be manipulated and separated with facility, and made of such material that they will not wear out by handling."[42]

This 5 × 2-inch format, closer to the golden mean than its predecessors, replaces the legacy of Croswell's wider paper slips (derived from the bound catalog format). While the old slips of paper aimed at presenting the relevant information in a single line, the new format spreads out over the surface. A colored line on the card separates head from body, and the body is printed with black lines, affording each detail its own line.[43] A wire runs straight through a hole in the cards, holding the bundle of entries together to protect them against unauthorized removal or accidental loss.[44] Wooden markers of the same format can be inserted at apt places, such as the transition from one letter group to another—described in the European tradition since Daniel Schwenter in 1636 as "noses" or "riders,"[45] since their upper edge peeks out above the card blocks, indicating the letter of the alphabet or the subject the cards are filed under. Looking back on the development of the apparatus, Abbot's assistant Charles Ami Cutter summarizes the interaction between hardware and software: "It was easier to plan the drawers in which the new catalogue should be kept than the system on which it should be constructed."[46] For this reason, the description of the catalog in the final report from 1864 contains an extensive explanation of the underlying principles of classification, a reference manual including detailed implementation instructions whose complexity exceeds that of the hardware by far.[47]

Before work on the catalog officially begins on October 22, 1861, Ezra Abbot carries out an extensive experimental phase to test the apparatus and the feasibility of the card-indexing process. He carefully measures the average data set: using this process, how many cards can be generated and recorded per hour? The result of 12.5 index cards per hour permits Abbot, in direct contrast to William Croswell, to calculate the production and construction of the catalog up to its temporary completion, and thus he can estimate how many employees are needed.[48] "Mr. Abbot and Mr. Cutter spend their summer vacation cataloging in the library," notes their supervisor, Harvard's librarian John Langdon Sibley, appreciatively.[49] Local library history usually portrays Sibley as a distrustful curator of the collection, concerned only with increasing and protecting the holdings and always prepared to discourage or reject potential readers.[50] On the other hand, it is Sibley who promotes the hiring of female employees, spurring progress considerably.[51] The number of employees steadily rises, and already in 1862 a part of the new catalog can be made accessible to the general public. Despite the generally appreciated advantages of the new apparatus, it takes another eight years until the catalog is finally completed in 1870, and the whole process comes to its conclusion.[52]

Regardless of its slow development, the exemplary new catalog at Harvard attracts attention even before its completion. In numerous descriptions—not just Abbot's own—its thoughtful principles are acknowledged, and the catalog consistently serves other large American libraries as a model.[53] Yet the sometimes sorry, sometimes glorious history of its development is supplanted once the use of card catalogs is no longer restricted to libraries. This major transition, initiated by a librarian in 1876, transforms the box of paper slips into the most important accounting and management tool in nearly every office by the turn of the century. This transformation of the "card catalog" into the "card index," this discursive transfer from *library* to *bureau*, is realized by a company called *Library Bureau*.

II Around 1900

6 Institutional Technology Transfer

Yet incessantly and annoyingly, the *perpetuum mobile* of business rattles and clatters amidst the cataloging and shelving work, which requires silence and concentration.
—Alois Jesinger, *Kataloge und Aufstellung der Wiener Universitätsbibliothek in ihrer geschichtlichen Entwicklung*

Reformation: Dewey's Three Blessings for America

On October 6, 1876, a 25-year-old assistant librarian signs the corporate charter of the American Library Association thus: "Number One, Melvil Dewey."[1] On Dewey's initiative, America's most famous librarians have assembled in order to found an association with the aim of promoting "the best reading for the greatest number, at the least cost."[2] Despite his youth, the initiator assumes the position of secretary, and from then on devotes himself to the development of American librarianship through the association.

From age sixteen, Dewey, who comes from a modest background and has been brought up strictly in the evangelical mind-set of white Anglo-Saxon Protestantism, is bent on reforming America in three ways. Already as a student, he rallies against alcohol and tobacco consumption, and for introducing the European metric system for weights and measures. Dissatisfied as a student, he declares the education system in New York State a failure: he feels it would be possible to learn double the amount in the same time, above all by simplifying English orthography.[3] Dewey begins by reforming his own name, first eliminating the superfluous letters "l" and "e" from his first name in 1875, and, four years later, by spelling his last name as well as everything else in phonetic transcription: "Dui, Melvil."[4]

In his quest for national reforms, Melvil Dewey is inspired by Edward Edwards's *Memoirs of Libraries*, an influential library history of his time. A struggle for "free libraries for every soul" becomes the third aim of his reform plan.[5] Within the scope of his service as an assistant librarian at Amherst College, Dewey has the chance to outline a more efficient organization of the library's routines and to optimize the library's management. Faithful to his motto, "My heart is open to anything that's either decimal or about libraries,"[6] he seizes the opportunity to combine two of his preoccupations, securing for himself a place in library history, and from 1930 onward a mark on every index card of the Library of Congress: the *Dewey decimal classification system*.[7] The design provides a hierarchy of fields of knowledge that, by means of a numerical code, can be differentiated again and again. The first decimal divides knowledge into classes from 0 to 9. Every further decimal serves to specify divisions of the preceding class (see also figures 7.1 and 7.2): "Select the main classes, not to exceed nine and represent each class by one of the (ten digits) nine significant verbatims. Subdivide each of these main heads into not more than nine subordinate classes, and represent each sub class by a digit in the first, or ten's, decimal place."[8] Once introduced to libraries worldwide, this system would grant unfailing and language-independent access to books,[9] an unambiguously composed numerical address thus leading directly to the desired text. "Lunar eclipse," for instance, is found at the fifth decimal place (0.52338) under natural sciences and mathematics (5), astronomy (52), descriptive astronomy (523), moon (5233). Furthermore, the classification satisfies four basic requirements for an ideal system as recorded in 1911 by another organization aiming for comprehensive access to world knowledge, propagated in Dewey's name; but more on this later. Dewey's decimal classification system also lays claim to features such as *boundless extensibility*, *general intelligibility*, and *clarity* to a rather unusual degree ("It must claim as little energy in use as possible"[10]). It partitions every field of knowledge down to the level of the individual component. "Thus, for instance, every lunar crater can be named unambiguously by further division of the number 52334."[11]

Given his preference for the decimal system, Dewey naturally starts a new job in Boston on April 10, 1876. He leaves Amherst College with three goals in mind: "adoption of the metric system, acceptance of simplified spelling, and the efficient operation of free public libraries properly

> **General Offices 6, 7 & 8, No. 32 Hawley Street, Boston.**
> Besides the ECONOMY Co., (a commercial corporation, manufacturing labor-saving devices for readers and writers,) three missionary educational societies have these offices as headquarters for the U. S. The collections illustrating each work are the largest known and are free to all. Each society desires as members all friends of education and progress, and many belong to all three. Otherwise they have no connection except the economy and convenience of offices in common.
>
> *"The best reading for the largest number, at the least expense."*
> **AMERICAN LIBRARY ASSOCIATION.**
> President, Justin Winsor, Librarian Harvard University.   Secretary, Melvil Dewey.
>
> *"The exclusive use of the International Decimal Weights and Measures."*
> **AMERICAN METRIC BUREAU.**
> President, F. A. P. Barnard, S.T.D., LL.D., Pres't Columbia College.   Secretary, Melvil Dewey.
>
> *"The simplification of English Orthografy."*
> **SPELLING REFORM ASSOCIATION.**
> President, F. A. March, LL.D., Lafayette College.   Secretary, Melvil Dewey.
>
> Each society has a SUPPLY DEPARTMENT, furnishing at low prices, often at half former rates, everything it recommends pertaining to the Library, Metric or Spelling Reform movements. Supported entirely by gifts and members' fees, the smallest sums are gratefully received.

Figure 6.1
Melvil Dewey's business card. (From Wiegand 1996, p. 62.)

stocked with good reading."[12] The recent graduate and passionate librarian embarks on his career as an educational reformer by founding three companies within one year: the American Metric Bureau, the Spelling Reform Association, and, toward the end of the year, the American Library Association, using capital borrowed from his landlady, his older brother, and a Boston publisher. For each of his companies, he opens a separate office. His business card, notably not printed on the back of a playing card, precisely summarizes all three businesses (figure 6.1). Each one is upheld by his conviction that time can be saved in nearly every business. This idea virtually becomes his obsession, the paradigmatic leitmotif of Dewey's life.[13]

No area of life escapes the dictate of saving time. Dewey's business accounting is subjected to his strategy of efficiency as well, yet with the consequence that it grows increasingly opaque. As indicated on his business card, not only does Dewey figure as secretary for each of the three new institutions, but the three offices are all listed under the same business address. Critical tax inspectors and accountants cannot help recognizing Dewey's unorthodox business practice of obscurely merging his companies, and his unorthodox accounting methods. Although he records even the smallest capital movement, the money circulates between institutions only *on paper*.[14] There is no actual transfer of the sums, preventing correct balances and consistent annual accounting. Calculation and storage must fail without that necessary transfer. According to demand, Dewey shifts the capital entrusted to him from one company to another—for instance, the expenses for an advertisement for his Metric Bureau in the Library Association's magazine, for which he acts as the publisher. Although he registers revenue in one account with a clear entry in the ledgers, he does not book it as a cost in the other account: there is only one account and one signatory, Melvil Dewey.[15]

Transfer: Library Bureau

Library Supplies
The starting point for the business history of card indexes is found in the small print on Dewey's business card (figure 6.1): "Each society has a Supply Department, furnishing at low prices, often at half former rates, everything it recommends pertaining to the Library, Metric or Spelling reform movements." The ardently pursued conversion to the metrical system that requires considerable new equipment—including rulers, scales, and instruments for measuring distances—is not all, though. According to Dewey, American library management around 1880 suffers significant organizational deficiencies. He is certain that as soon as his three reforms are established, new office tools will be needed to serve the new paradigm. His "Supply Departments" are designed to meet this anticipated shortage of suitable materials. His library supply department will "serve as a guide in selecting the best forms of the various library supplies."[16] The necessary cataloging tools are custom-made and delivered at a discount, "in such a way that they can be used for the catalogues of all libraries."[17] Dewey is convinced that library efficiency can be increased only with uniform cataloging techniques on uniform materials of consistently high quality. In his view, internal business processes in library management and reader service, above all the catalogs, will need to be submitted to strict standardization procedures. With the help of generalized forms, rules, and systematic arrangements for everything from centralized and standardized printing of index cards, to drawers and boxes to fit them in, to inkwells and pens, he begins to "synchronize" libraries. Application of all these innovations is to lead, in Dewey's vision, to "an immense saving of time."[18]

Standardization

Thus, one of the first projects undertaken by the *new* American Library Association in January 1877 is a standardization program, with the aim of establishing rules for cataloging processes, catalog entries, and bibliographical norms (in the best tradition, yet without knowledge of the algorithms of the Josephinian project).[19] One issue for this bibliographical *system* is the external appearance of book catalogs, particularly the question of correct measurements for a *Standard Catalogue Card*.[20] As secretary and host of the meetings of the *Co-operation Committee,* Dewey invests less into originality than into realizing his reform principle of optimization. He by no means invents the card index anew, yet he aims to improve Ezra Abbot's catalog construction manual, finally patenting the entire object and touting it as an innovative achievement meant for permanent application. "The first catalog cases were made with two rows of cards in a drawer. Some of these cases after forty-five years are still in use."[21] The initial drawers—which, complying with Moser's construction manual, hold 1,000 paper slips in two rows (see figure 6.2)—are followed by numerous products with widely varied features. The otherwise voluble minutes of proceedings withhold any hints as to existing concepts or established paper slip techniques

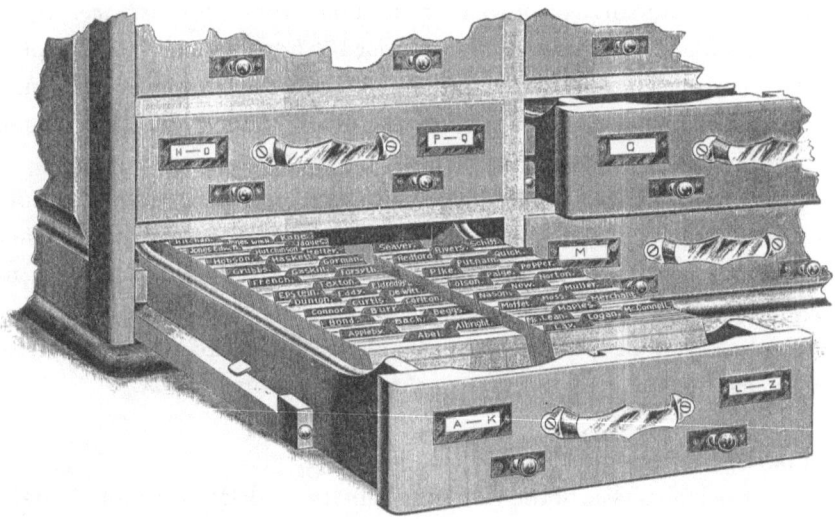

Figure 6.2
Two rows of paper slips in a drawer of a Library Bureau box, ~1894. (From Davidson and Parker 1894, p. 62.)

that may have served the committee as models. Instead, they aim to create the sense of a homegrown American technology. European magazines are slow to defy this claim, and do so rather timidly. "I should not like to enter upon an inquiry on the origin of the Card catalogue. One thing appears certain; its origin is not American, as seems to be generally supposed."[22]

However, before all the index cards settle into their drawers, there remains the issue of standardizing the heterogeneous. The American Library Association helps by gaining influence in important libraries. As the association lacks competition, it can easily prescribe standards for the young and disorganized market. The Co-operation Committee, expressly founded for this purpose, suggests a standard format for catalog cards. Cards with the dimensions 5 × 12.5 cm, as Dewey's *Supply Department* already sells them, come into use, as well as cards in American postcard format.[23] One year later, though, this format is declined for global usage by the Universal Postal Union, which standardizes the postcard worldwide as 9 × 14 cm.[24] At its second conference early in September 1877, the American Library Association votes to follow the committee's proposals, and the standard dimensions for index cards from then on correspond to the two most frequent formats in America (compare nos. 12 and 14 in figure 6.3 to all European formats). Although there are sporadic proposals for an internationally recognized card catalog format—for instance, *Professor Burchard's internationally uniform catalog slip* from Vienna (1880), explicitly referring to global postcard formats—in 1908 the Institut International de Bibliographie decides on an internationally valid format for catalog cards: 7.5 × 12.5 cm, the measurements of the American postcard.[25]

Thus, the American library community not only defines two generally accepted formats, but also prevents diversification and occasional library debate over what the right format might be. Meanwhile, librarians in Germany engage in extensive and passionate debates over format questions after 1908, in spite of the new worldwide standard. "I cannot praise as an advance the eyesore they call this international format."[26]

Corporate Genealogy

The first business year of the American Library Association Supply Department yields a profit of only $300. The department develops slowly, until in March 1879 Melvil Dewey separates it from the Library Association (not least to counter the denigration of his products as "gimcracks" by some of

Institutional Technology Transfer

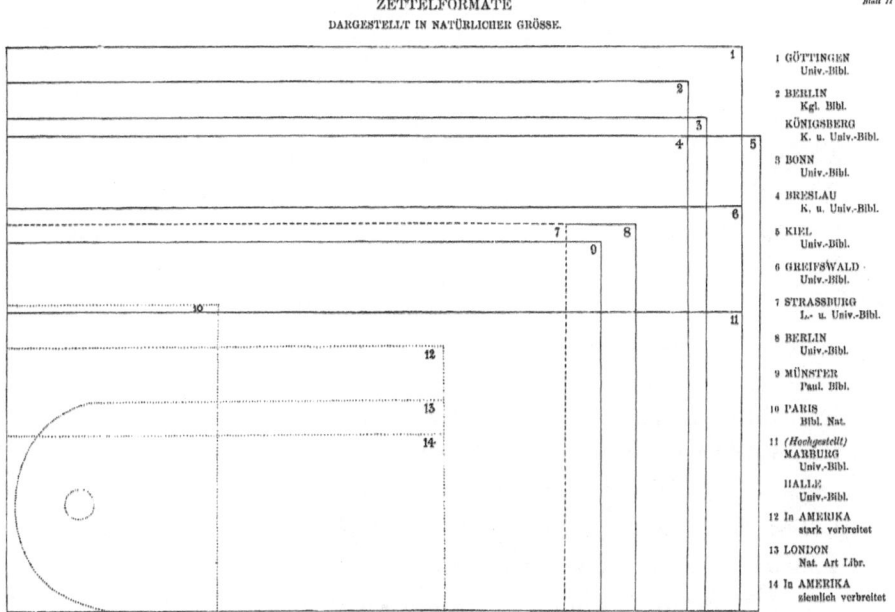

Figure 6.3
Diversification of paper slip formats reduced to 65 percent. (From Milkau 1898, appendix, p. II.)

his associates), reorganizing the company with the aid of his new bride's dowry, as well as additional capital infusions from his former landlady, and calling it *Readers and Writers Economy Company*.[27] Business improves, and rapid expansion ensues, necessitating larger offices. By December, the company is registered as a shareholder company, with the mission to operate as the sole manufacturer of about 400 office supply products (including desks, study materials, and library supplies—items like the "Reader's Readyrest Perfect Vacuum Inkstand" or the "Economy Eye Shade") and to open new markets. Everything remains dedicated to the principle of helping people to save time, money, and effort. Dewey signs approximately 120 contracts with producers of office equipment and arranges bulk purchases to resell via the ever more numerous branches of his Readers and Writers Economy Company.[28]

However, rapid growth leads to difficulties in management, still consisting of Melvil Dewey alone. He never ceases to shift between the various positions and salaries in the institutions he serves as a secretary, president, publisher of various magazines, and so forth. In fall 1880, his negligent, hopelessly tangled bookkeeping backfires. The executive board of the Readers and Writers Economy Company examines his books and finds completely inscrutable accounting, deciding immediately to initiate an external investigation, as well as issuing an injunction against Dewey. The chaos, once officially confirmed, puts Dewey in a bind: he is forced to resign as manager and president of the company, and other demissions follow. The executive board limits his financial authority, and finally under massive pressure he leaves the company, having lost the trust of his remaining directors—but stubbornly defending his aims and reformist intentions as purely altruistic.[29]

Notwithstanding, Dewey does not abstain from his desire to supply the library world with materials.[30] While the Readers and Writers Economy Company now starts to pursue its own profitable paths, already on March 20, 1881, Dewey announces the foundation of a new company—which is simply his old one, cleansed of accounting discrepancies, once more bankrolled with borrowed capital and unequivocal messages of continuity: "All Metric articles will be sold as before under the name Metric Bureau [...] The Library Supply Department will go under the name *Library Bureau*."[31]

The rejuvenated company could not have been named more aptly. The office in the old rooms of 32 Hawley Street in Boston serves its company founder not only as administrative residence for his usual projects and shops, but above all as an exact declaration of the purpose of how to go to market with his products. As of spring 1881, Library Bureau points the way to a significant technology transfer between library and office, turning library-technical discourse into a business.[32]

Despite his sudden departure from the Readers and Writers Economy Company, Dewey devotes himself with unbroken zeal and unfettered energy to the new company, necessarily disregarding his other proposed reforms in the meantime. So as not to endanger his larger reforms for too long in favor of trading in library supplies, in 1883 he puts Herbert E. Davidson in charge of Library Bureau, but not before ensuring his own veto power against anything that might diminish its benefit for libraries. At first, the businesses fail to thrive under Davidson: Dewey still exerts considerable influence on questions of business strategy, insisting on relatively low margins as an incentive for libraries, once again leading the company increasingly into financial straits. Despite Davidson's excellent commercial instincts that make Library Bureau prosper by developing a network of contacts with other enterprises, he lacks a talent for accounting. As a partner in the corporation, he uses his shares for private expenditures and finally becomes unworthy of credit.

Although Dewey does all he can to prevent this renewed crisis from attracting public attention, the *Library Journal* of the American Library Association reports in its March 1888 issue on "The Library Bureau Failure."[33] When the books are balanced, the creditors demand $27,926.98. Dewey has no choice but to buy out the entire company. "My life work involves having a succesful LB [Library Bureau]... & I must make sum sacrifices."[34]

The Transfer

Bearing considerable financial losses, Dewey succeeds in averting the ruin of Library Bureau. On May 26, 1888, the reorganized company meets under the presidency of Melvil Dewey. Despite his role in the previous bankruptcy, Herbert Davidson is again employed by the company, but this time to concentrate less on financial management than on "manufacturing and sales."[35] E. W. Sherman, employed as an accountant since 1887 and later the woman to lead the Indexing Department, insists on an organizational change. She has come to appreciate the ease with which card indexes allow system maintenance and persuades Dewey to let her transfer her accounting books to cards. Yet Dewey, wedded to the traditional library paradigm and marked by his experiences with untransparent accounting, remains distrustful of loose paper slips, fearing data loss and misplaced entries. "The inventor of the card system exclaimed: This won't do! This won't do! Why? I asked. Because cards will be lost, he replied."[36] Nevertheless, Sherman insists and negotiates a six-month trial. She assures Dewey that if only one card is misplaced, they will transfer the whole process back into a bound book, at her own expense. The experiment succeeds, and even begets a business idea. It demonstrates that accounting on slips of paper is clearer, easier to audit, and above all quicker to handle. From then on, Library Bureau administers all the company's accounts by means of the *card index*.[37] Looking back on the early days of the company, an internally produced 1909 Festschrift summed up the momentous idea, which the manager Davidson had taken from his secretary Sherman. With unconcealed pride, the company proclaims the invention of a worldwide *new science*:

It was in 1888 that Mr. Davidson conceived the idea, by no means startling in itself, which opened up the real future of Library Bureau. This idea was that the card catalog, then in fairly general use by libraries, could be adapted with advantage to certain *commercial indexes*.

Mr. Davidson had no exaggerated opinion of the value of his idea. He thought that if it worked out successfully it would add a new item of profit to their small business. As a matter of fact, Mr. Davidson had discovered a great new science—a science whose principles underlie every department of commercial activity—a science which, through the medium of the card system, has revolutionized business methods in the last twenty years, and which, year by year, is playing a larger part, not only in the commerce of this country, but of the world,—the new Science of Business System.[38]

The successful introduction of commercial data based on paper slips begins with the most traditional note bank in Europe. In 1852, the Bank of England rearranges its account management on index cards stored in boxes.[39] This prompts numerous banks to follow suit; the first one in America is the Williamsburg Savings Bank in New York in 1884, after its president was inspired by the library catalog at Columbia University— precisely at the time of Dewey's management of this institution—as a potential tool for the managing of bank deposits.[40] Meanwhile, Herbert Davidson develops a far-reaching network of contacts as sales representative of Library Bureau, thus spreading his experiences with the card index among his business partners. Insurance companies show special interest in Library Bureau supplies after Davidson arranges card indexes for New York Life, Equitable Life, and other companies: in addition to drawers, he regularly delivers index cards printed with the data of those who have requested life insurance at the respective companies, but were considered high risk and refused. This information exchange pivots on central collection in Library Bureau, and daily replies return from the company's printing press to all subscribing companies on neatly formatted index cards in impeccable alphabetical order. The expected gain in money and time is immediately confirmed, and in short order, nearly all the life insurance companies in the United States unite into an exclusive card index and mailing list club, potentially refusing the admittance of new members.[41] This special insurance policy against unsatisfactory insurance policies not only becomes a best seller, but also proves how effectively central catalogs and index card sets can be shared and used. For the idea of the central collection and typographical processing of modular library information on cards is what

Library Bureau had been trying to launch in conjunction with the Library Association since 1878 for all American libraries large and small—always failing because of the members' unwillingness to commit.[42]

News of the time-saving index card method travels fast in East Coast business circles. It soon reaches the banks: Emigrant Savings Bank turns to the insurance company Equitable Life, already equipped with paper slip boxes, and inquires about where to buy card indexes. Equitable Life refers the bank to Herbert Davidson of Library Bureau with the remark that "he knew more about card indexes than anybody they could go to."[43] By 1890, Library Bureau receives the order from Emigrant Savings Bank and proceeds to open an Indexing Department, devoted exclusively to selling card indexes to industry. In the same year, more large-scale orders from various banks and prestigious companies follow, including Greenwich Savings Bank, Massachusetts General Hospital, and the Carnegie Steel Company. The indexes pass their tests to the satisfaction of the ever more efficiently organized managers. No area of data processing remains untouched by card indexes. In 1892, Herbert Davidson signs the first contract for the organization of a correspondence recording system. "[W]hen it was reported...that the work was completed, Mr. Krauth of the McConway & Torley Co. asked Miss Pope to bring him a certain letter which she promptly did. 'There,' said Mr. Krauth, 'it was to find this letter that I have had this correspondence indexed. It is worth tens of thousands of dollars to us.'"[44] Unless a letter is purloined, from now on it will never fail to reach its destination.

This unexpected and rather haphazard clientele constantly purchases more card indexes. The company receives many orders that require hardly any investment. Finally, the offices of Library Bureau consolidate and achieve considerable prosperity, to the satisfaction of its partners and shareholders. Only Melvil Dewey expresses his suspicion: while welcoming the yields of his major share in the business, he still wishes to offer high quality and low prices for the benefit of libraries, not for profit.[45] With the help of his veto power and control over his majority shares, Dewey opposes the increasing commercial expansion planned by Davidson and his managers. Their aim to control the quickly growing market runs counter to Dewey's founding idea of the paramount interest of public libraries.[46]

In the midst of the general economic crisis of 1893, Library Bureau manages to triple its working capital. Annual revenues rise from $34,919.92 in 1889 to $193,185.56 four years later, "most of which came from catalog card stock, cabinets, and trays sold to businesses, not to libraries."[47] As a result, President Dewey negotiates a new contract with the executive board, securing 20 percent of company profits as well as control over development for himself.[48] These dividends prove profitable in the persistent upward trend in the market for office equipment: while in 1894 Dewey earns $3,650, by 1895 his profit has increased to $5,750. The next years bring even more rapid growth as the boom continues. The revenue growth of Library Bureau owes its reassuring ascent to standardized slips of paper, wooden drawers, and boxes sold at fair prices. Credit for the substantial commercial success of card indexes goes to Herbert Davidson. Besides banks and insurance companies, he wins railroad companies, government departments, and other large-scale enterprises as loyal customers.[49] Against Dewey's attempts to fetter the capitalist expansion of Library Bureau, starting in 1891 management opens numerous offices and sales branches with demonstration sites in other towns to promote future business and enhance its market power in other areas.

It is worth noting that at first, no explicit advertising seems necessary for the development of new markets or the acquisition of customers in the form of announcements or brochures, as word of the favorable qualities and promising achievements of the card index spreads in business circles.[50] Once the market is contested by other office supply companies and new competitors, Library Bureau also buys advertisements, usually in the form of magazines and order catalogs intended for businesses. In the product catalog of 1891, the advertising slogans still address librarians above all, praising new trends and touting incentives for a changing of the guard: "There is hardly a library article on our list that is not also used in offices, so that the Bureau, beside its mission of representing the focalized experience of the libraries, is finding a larger and equally interested clientage in wide-awake, energetic business men and institutions."[51] Libraries seem to lack this energy for the time being.

Showing the same reserve with which they met the index card printing offered as a service by Library Bureau since 1878, the management of a majority of American libraries reacts cautiously to Library Bureau's steady praise of the card index. Only after considerable delay does the idea return to the libraries around 1900, finally asserting itself as their very own technology for data management with unlimited and unbounded use. Meanwhile, the management of Library Bureau has started to turn

Institutional Technology Transfer 99

away from librarians as its desired customer group, and instead stokes the interest of commercial companies with ever new and improved products. These companies have been quicker to adapt their bookkeeping to mobile and user-friendly catalog systems, saving half the previous time and thus adhering to the paradigm of widespread time-saving. The technology transfer of the card index becomes explicit. "All the features that make it the greatest library invention apply equally to commercial lists."[52] The claim to universality gradually steals into the encomiums praising the card index as an all-embracing and infinitely applicable machine: "From an author's catalog it has spread to an almost *infinite application*. Every list, record, index, *etc.*, that is in a *state of growth* can be thus kept with great saving of labor. Business houses find it invaluable for lists of goods, customers, discounts, and the *1,000 growing* records of commerce. Science adopts it even more widely, and its use is spreading with *growing rapidity*."[53]

Unabated growth combined with a belief in limitless progress is an offer bound to convince interested company managers. These advances must be accompanied by a storage system that can meet unexpected developments and allow infinite expansion. In order to appeal to every businessman, Library Bureau places advertisements with this verbose, tailor-made analogy:

> Everyone who handles large lists of addresses or keeps in convenient order miscellaneous facts knows the great difficulties involved. Ingenious devices have been invented, and wonderful computations made as to the possibilities of alfabetizing, some involving the close classification of a million names from directories. Records and indexes have been based on the results. But fitting these calculations to some prospective list is like making a suit of men's clothes for a ten-year-old boy in anticipation of his growing to the form and size prescribed. He fails to do so. Arms protrude, space is in the wrong place, the coat won't button; and yet the chances of a good fit were better than that needs will agree with prescribed space in these inexpansive, arbitrary indexes. Libraries recognized this difficulty in their work years before the great and growing records of commerce, invented the card index, and discarded blank books for their indexes. Until recently its use has been confined to them, but somehow with the customary avidity of commercial life business men caught the idea, and without the influence of active propaganda have to a wide extent adopted it.[54]

The unexpected and rapid spread of library card indexes in business offices influences the budding organizational discourse and management's attention begins to shift: Library Bureau knows how to sound out new markets, and how to harness feedback to engage new business segments. "The Bureau is rapidly extending its work still more on the business side, and hundreds of the most successful and extensive corporations in the country have given testimony that they have profited greatly by adopting the methods and devices which have done so much in the past few years toward making money spent on free libraries accomplish a greater work."[55] From the successfully occupied central position between the coordinates "library" and "office," management shifts gradually toward the second term to define the office realm as a new and very fertile market. The benefit of libraries remains no more than a (nearly) forgotten starting point that must now give way to a focus on the office.[56]

Product / System / Manufacturing

In 1895, for the first time, Library Bureau invests heavily—as the company catalogs repeatedly emphasize—in its own factory to actually manufacture the index cards it had turned into a bookkeeping tool.[57] Quality paper and matching cards are the decisive criteria for guaranteeing long-term use. "It is important that all cards used for catalogs or indexes should be exact in size, as the slightest variation interferes with facility in handling them. A low card between two higher ones is bridged by the fingers and lost."[58] After the successful introduction of the American Library Association's card format, the attempt to standardize index card quality once again seems to be accepted by the market. Herbert Davidson is delighted about the successful word of mouth and the growing sales. "We have made a standard never known to the paper trade before."[59]

Starting with that "greatest library invention," the card index, Library Bureau expands its product range bit by bit into areas that are based on a library or archival cataloging logic, yet generate massive economic demands. In 1869, two different file cabinets are invented: on the one hand, E. W. Woodruff's wooden containers for folded documents, used mainly by the federal government; on the other hand, the so-called *vertical files* that allow folders to be stored in an upright position:[60] the Amberg File and Index Company manufactures a container to keep the written material unfolded and on a slant. As a special order for Nathaniel S. Rosenau, employee of a charity organization in Buffalo, in 1892 Library Bureau transfers the idea of this special construction for the storage of files to an open briefcase with compartments for the vertical storage of sheets of paper. Rosenau displays this system at the 1893 World's Fair in Chicago, where it attracts the attention of an employee of the local Library Bureau branch. As the idea oscillates between clients and Library Bureau, anonymous employees add the modular extendable device to the company's regular product range. Under the name *vertical filing cabinet* or *suspension file cabinet*, this apparatus more than doubles the company's revenues after 1897.[61]

By 1918, a third of the whole product range consists of registry equipment, with a clear preference for the vertical filing cabinet. With the emphatic victory of card indexes in the most liberal of all market economies, government authorities in the United States also become increasingly interested in this form of data management. In 1897, Library Bureau receives an order to install Bertillon's system of anthropometric list processing on an index card basis for the National Bureau of Identification; this is enhanced in 1905 by a fingerprint card index as a national police database for the identification of criminals.[62] Other government agencies like the New York State Prison Commission follow the example and ask Library Bureau to equip their administrations with office management and accounting systems based on index cards.[63] The demand from government and business for efficient reference management—that is, sorting and filing papers—is ultimately a reaction to the recent *paper flood* catalyzed by stenography, carbon paper, and typewriters.

After government institutions dealing with statistics in the widest sense begin to pay attention to Library Bureau, it is hardly surprising that Herman Hollerith, who has been in touch with Herbert Davidson since 1895, does the same. On March 31, 1896, Library Bureau and Hollerith's Tabulating Machine Company (later to be known as IBM) sign a (mere) three-year contract. Their goal is to join forces in order to support censuses abroad, railroad and insurance companies, and so on. Ten percent of the net profit goes to Library Bureau, which provides the cards required by the tabulating machines. As a consequence of this short period of cooperation, Library Bureau opens a branch in Paris to assist the French Ministry of the Interior in administering its census. However, other projects soon lead to differences between the partners. Hollerith claims control of the entire transportation market; Davidson disagrees. In March 1899, their collaboration comes to an abrupt end.[64]

In 1909, Library Bureau controls more than 10 factories, 32 agent offices in American and European cities, and 3,000 office workers. A small department for library supplies overcomes financial plights and turns into a thriving corporation, opening a hitherto virtually nonexistent market. With the aid of its own manufacturing, offices are equipped, management is optimized, companies are analyzed. Library Bureau, though, never forgets where it came from. Herbert Davidson, to whose outgoing sales talent the company owes its position between "library" and "office," proudly states in a confidential report, "The two great systems which have revolutionized accounting and record keeping; viz., the card system and the loose leaf system, had their origin in library work [...] The enormous growth of the Library Bureau has come from the development and application of these library systems and methods to commercial work."[65]

Digression: Foreign Laurels

Although the management of Library Bureau appears fully aware of the discursive and practical origins of card index systems, the exact origins and first applications are still not made explicit. For reasons of market power alone, and faced with increasing competition, it seems opportune to hide credit other than one's own, and to credit oneself with the independent invention of data management systems. "L B [Library Bureau] has been claiming for 30 years to be the inventor of this system [card index]."[66] However, this belated criticism of the company's advertising strategy fails to take hold. Throughout his entrepreneurial career, Melvil Dewey has presented himself as the inventor of this time-saving concept. Only after 1900, when Dewey has already withdrawn from active management of Library Bureau to devote himself to his retirement resort in Lake Placid, does James Duff Brown's European criticism cast some doubt on Dewey's claims.[67]

With the wisdom of age, Dewey grants that there might have been a limited American prehistory. Reacting to requests to discuss details of the origin of data management, he begins to revise his rather ostentatious claims.[68] In a letter, Dewey recounts—in his peculiar orthography, and still assuming a protagonist's role:

The erly story of the card system is known to very few. Attention was cald to this by C C Jewett at the 1st convention of librarians in the world, held in N Y Sept 15 to 17, 1853. Jewett used ½ a sheet of foolscap, but the idea was there. Harvard Colej and other universities gradually adopted the plan for their card catalogs. I got the idea from Jewett, and Howard used it for the card catalog of Amherst Colej library of which I had charj in 1873 to 76. I also introdust the loose leaf ledjer idea.[69]

His response to an earlier inquiry was more nebulous and a bit less unambiguous. Dewey's correspondent, Burt C. Wilder, who himself lays claim to the same invention, is forced to insist:

One query more as to the slips. You say they had been "recommended in 1851 [sic] at the first library conference of the world in N Y City, and Ezra Abbott [sic] began using them about that time in the Harvard College library." Who recommended them? Where is it recorded? I was at Harvard from 1859 to 1862, but do not recall them. When I began to use them in 1867, I was not conscious of any precedent, and my communication to the Boston Soc. Nat. History does not intimate any. A few years later (my diaries will tell but they are at my summer home) I spent a summer on a card catalog of the brains and embryos at Cornell; the general use of the slips was much accelerated there by their systematic employment by the late G.C. Caldwell, who got the idea from me; I am not sure but what my part in the introduction of slips may be regarded as my most important contribution to human progress, so I may be excused for a wish to learn the facts.[70]

In 1862, the strict separation of public and library-internal access to paper slips still prevailed at Harvard. However, as already described, this would soon change with Ezra Abbot's opening of the card index to the general readership. Dewey soon generously responds to Wilder's inquiry: "Abbott [sic] used the cards only in the library. I did not begin spreading their use until 72-3. Your work in 67 is the earliest I have heard of in pushing them for personal use. You are right in thinking it something to be proud of. I am using the 3 inclosed slips and urging others as the most compact and logical way to communicate."[71] Despite the long European tradition, and despite the long-established albeit relatively young American genealogy, and counter to the achievements of Herbert Davidson in transferring library technology to office organization, Dewey, for the rest of his life and against his better judgment, refused to admit that it was not he who played the chief part in the index card game.

In fact, it was Charles Coffin Jewett who, at the first worldwide librarian's conference in 1853 in New York (which Melvil Dewey naturally did not attend, being only two years old at the time), produced his card-based catalog. Jewett's plan (as in revolutionary France half a century earlier) was to collect all U.S. bibliographical information in one location—namely, the

Smithsonian Institution in Washington, D.C.—compiling data on the basis of submitted catalogs whose entries were to be transferred onto stereotyped clay printing plates, with one plate per text, and arranged in alphabetical order in a central memory bank on "sliding shelves" or "shallow drawers."[72] Like a type case, this supply was to form the prototype of a standardized comprehensive catalog. As soon as one of the participating libraries planned to print a new catalog, the corresponding plates would have been selected from the collection and printed in a folio. This idea drew harsh criticism from many sides, though. Despite its advantages of uniformity, flexibility in the arrangement of entries, and consistent up-to-dateness (similar to the index card principle), the plan was mocked as Jewett's "mud" catalog and never realized.[73]

Industry Strategy

The marketing strategy of Library Bureau relies on standardized paper strips based on unmentioned Old World library technology, protected early on by American patents. In the early stages, bulk purchasing of conventional office supplies of "good quality" serves as the foundation of a product range selected by library experts and then resold under the brand *Library Bureau*. Yet after 1888, Dewey is forced to restructure the company once again, incorporating it as a shareholder corporation and changing his business strategy as well. His emphasis is now on self-sufficiency, speeding up product development, and manufacturing the products himself. New developments are usually registered immediately at the U.S. Patent Office, a good indicator for the productivity of the research and development departments. Between 1893 and 1916, Library Bureau claims nineteen patents for the *card-indexing system* alone, mostly for rather small improvements of the concept.[74] None other than Herbert Davidson lays claim to the influential idea of a protective device, piercing holes into the cards to guard them against theft and dispersal.[75] The first punched-card index features only one control bit for its own protection. (The last bit in the life cycle and the era of the index card involves its transfer, more precisely its dispersal: for instance, in a competition where cards referring to books are fastened to balloons and compete for the farthest trip. Or transfer by projectile: in a graduation ceremony at Cosumnes River College in California, a librarian pulls a gun, shoots, and destroys punch card after punch card.[76]

Institutional Technology Transfer

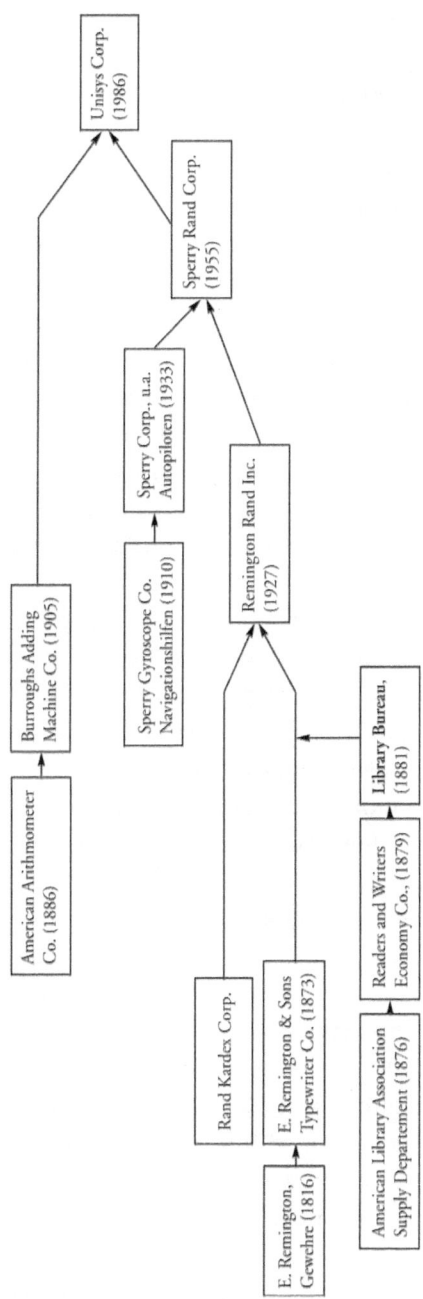

Figure 6.4
Diagram of mergers and acquisitions

Here the slips of paper act not as a medium, but as a receiver. Their holes are for good. Legend has it that the sight of a ticket-punching conductor gave Herman Hollerith the idea to store information by means of holes.[77])

Although Library Bureau can claim to be the first business to provide library supplies along with other office machines and materials, it is soon faced with competition. Shortly after the blossoming sales of card indexes establish format standards, manufacturers like Art Metal Construction Company (founded in 1884) and Yawmen & Erbe of Rochester, New York (founded in 1898) show up in the market. Globe Wernicke (founded in 1882) manages to exceed Library Bureau's revenues within a few years, although the same fate befalls them both. In 1926, none other than E. Remington & Sons Typewriter Company, spun off by gun producer Eliphalet Remington II in 1886, acquires both Library Bureau and Globe Wernicke, merging them the following year with Rand Kardex to form Remington Rand Inc.[78] A department called Remington Kardex Bureau spurs the decisive advancement of the index card to an automated storage device principle whose origins refer once more to Europe, that is, to the eighteenth century and Jacques de Vaucanson as well as Joseph Marie Jacquard's punch cards.[79] After 1958, five electric-pneumatically linked Remington Rand typewriters print the paper slips of the last analog catalog of the Austrian National Library, five copies synchronized by compressed air. However, they prove inferior to the more robust and soon widely used electric typewriters of the International Business Machines Corporation, and as a result are disposed of.[80]

The intertwined genealogy of card index makers and typewriter manufacturers, leading to the production of the universal discrete machine, remains an American history of mergers and acquisitions. Remington Rand Inc. (manufacturer of UNIVAC in 1957) asserts itself as a small but persisting line of competition to IBM, merges in 1955 with Sperry Corp. into the Sperry Rand Corp., which in turn is acquired in 1986 by William Seward Burroughs's Adding Machine Co. and is combined into the Unisys Corp. (figure 6.4). The appreciation of corporate history exhibited by employees of the Sperry Rand Corp. is praiseworthy enough: in 1976, a small independent company is spun off from the office supplies department threatened by liquidation, going by the name *Library Bureau* and to this day engaging in a modest yet flourishing library furniture business in Herkimer, New York.[81]

7 Transatlantic Technology Transfer

The principles of the Library Bureau equipment are currently conquering the world.
—Paul Ladewig, *Politik der Bücherei*

Throughout the nineteenth century, the practice of using index cards not occasionally but permanently in cataloging large book collections spreads among European libraries, finding its way across the Atlantic into the New World, to be naturalized after 1861 as a new technique accessible to a general audience. Ezra Abbot could be said to have sponsored its green card. Already in 1877, one year after its foundation, a delegation from the American Library Association departs for Europe to found a subsidiary in the United Kingdom. Melvil Dewey and his future wife are among the travelers; they meet during idle hours aboard, bonding over card games.[1] The delegation succeeds in establishing the British Library Association, which remains closely linked to its American parent. Yet the index card principle is recognized much more slowly in European libraries, so that the American delegation's suggestion to dismiss traditional bound catalogs in favor of the reimportation of loose card catalogs—insofar as the group aspires to such missionary efforts—is met with resistance. Though catalogs are by then accepted as universal search engines, they remain fixed in the library paradigm of the bound *book*.[2] Library employees continue listing new entries systematically in bound catalogs and their supplemental volumes. The following part of this study aims less at reconstructing the extensive and passionate debate that finally yields to the triumph of the card index over bound catalogs[3] than at sketching the dissemination of the American indexing technique in the Old World. Largely stripped of its provenance, the card index *reenters* Europe via two channels. On the one hand, the American business community vouches for the latest office machines and above all the card index as a time- and labor-saving device, transferring the early scientific management ideas of Frederick W. Taylor and Frank B. Gilbreth into the office.[4] On the other hand, a massive European demand for new library technologies develops independently of the office innovation realm, owing to the construction and renovation of large national libraries.

Supplying Library Supplies

The Library *Ge-stell*[5]

From a library perspective, the second half of the nineteenth century is an era of construction. In 1854, a new phase of library architecture begins with the extension of the British Museum. Here, the architectural modern age meets growing storage needs by means of building stacks that no longer rely upon representative shelving, but obey the form dictated by shortage of space and the need for optimal access. In 1859, the Bibliothèque Nationale in Paris follows the principles established in London, lowering ceiling heights and furnishing shelves that no longer hug the walls, but "stretch out like feelers into the middle of the space."[6] This French-English model dominates the designs of a first wave of enlarged library buildings in Germany. The administrators of university or state libraries in Rostock (1866), Karlsruhe, Stuttgart, Halle, and Göttingen (1878) are convinced by the new paradigm, even before construction technology between 1890 and 1915 allows buildings to become one big bookshelf.[7] The system proposed by a Strasbourg locksmith named Lipmann is first applied in Marburg, then improved in Gießen and Tübingen before eventually being installed in the imposing new Royal Library in Berlin, *Unter den Linden*.[8] Lipmann's concept consists of a multistoried bookshelf resting on iron abutments. "The essence of modern technology lies in enframing."[9] The individual shelves can be hung according to the formats of books, and like false ceilings they can be placed flexibly in prefabricated recesses. Librarians need neither ladders nor detours to retrieve books from the collection: "Enframing means the gathering together of the setting-upon that sets upon man, i.e., challenges him forth, to reveal the actual, in the mode of ordering, as standing-reserve."[10]

The cramped situation in the largest library in Berlin comes to its long-awaited end with the inauguration of the new building by His Majesty,

Emperor Wilhelm II, on his birthday in 1914. In the old buildings near the Opera, complaints about the shortage of space had accumulated, not least with respect to the catalog room: "Between two large cabinets in front of an admittedly wide interior window overlooking the courtyard stood, very cramped indeed, no less than nine worktables that badly obstructed access to the index card boxes. Anyone who needed to tend the index cards here had to drive the people out of their seats, often for hours; one employee assured me plausibly that he devoted himself to this unenviable work on Sundays, in order to be undisturbed and not to disturb others. That is the true Prussian spirit."[11] Unfortunately, the index card boxes in the Berlin library only briefly enjoy room to expand. "Even in the new building, they filled an entire room [...] and it required extensive calculations to accommodate the card index generally in a tolerable way in the gigantic new hall. At this point, to install even one more cabinet would be impossible."[12]

Once finished, the extensions and new buildings require adequate interior architecture appropriate for library technology, and the question is, which system should be applied. Whose shelving concept wins? What brand of index card boxes should be purchased? In 1912, Dr. Paul Schwenke (1853–1921), the first director of the Royal Library, visits the United States for inspiration, and to examine the library equipment there.[13] It is no coincidence that a programmatic monograph called *Politics of the Library* comes out in print that same year, drawing an explicit library connection between economic and organizational discourse. In it, Dr. Paul Ladewig (1858–1940), who is to become the director of the Central Institute for Education and Pedagogy *(Zentralinstitut für Erziehung und Unterricht)* in Berlin, associates factories and libraries structurally to stimulate the infusion of commercial administration into library administration. Under the heading "standardization," he compares the machinery in a large factory to the "apparatus of the library," whose productive success is based on a "standardization of the individual parts and precision of the work."[14] Ladewig, convinced by the company's publicity leaflets, recognizes Library Bureau as the pioneer of library supplies. Thus, in his view the only correct library policy is to equip one's own library with Library Bureau materials: Ladewig recommends them enthusiastically and ends up making a case for adopting American norms:

In any large library in Germany, one tends to find ill-fitting yet expensive cabinets that remain open and accumulate dust, and in smaller German libraries one occasionally even finds tattered cardboard boxes. Catalog slips of the normal format might be delivered from the bookbinder, but even for bookbinders, precision cutting is a difficult task; tiny differences from one delivery to the next lead to cards standing unevenly in the apparatus, slowing down the work even if enough header cards are deployed, and heavily used catalogs rapidly wear out. The catalog slips must be as precise as artillery, which can be achieved only in the framework of large-scale production cutting several million cards per year, and that does not yet exist. The catalog slips as delivered must remain uniform in quality and thickness over decades. This is a crucial point which we cannot emphasize enough. Precision and productivity absolutely depends on precise materials. One must have learned and viewed plenty of things before one can act independently following the preliminary work of Library Bureau.[15]

A dark threat. What does "plenty of things" possibly refer to? Does an enterprise need to fail twice—as in Dewey's case—before it opens new markets or manages to consolidate its accounting? Or is it necessary to take Clausewitz's line that "war is merely a continuation of politics by other means" seriously? Is 1912 possibly an anticipation of an impending wartime economy? If the management of soldiers and the management of books in Vienna around 1800 converge, index cards a century later are substitutes mediating between reader and book. Made precisely and according to standards, they in fact become ammunition in the shape of precise media of transmission, that is, projectiles.[16] As a former librarian of the Essen Library as well as the Krupp corporate libraries (and at times a welcome guest at the Krupp residence), Paul Ladewig is aware of the distributive power of his library policy book, which went into three printings, undertaking as *normalization* nothing other than the recommendation of "impartially working" Library Bureau products.[17] The unequivocal program of the industry-friendly librarian barely comes as a surprise: "the strict application of indexing practices in specialized publications, [...] an international format of index cards, a coherent collection of office forms, the construction of storage towers for libraries, [...] taking precautions against the effects of bombing in the likely event of war" are among his innovations in the field of German librarianship.[18]

Punch Cards

The Viennese card index thus manages to assert itself as the permanent basis for book search, and this expansion of interfaces requires a gradual yet inevitable extension in everyday interaction with these delicate devices.

What Ezra Abbot initiated for American libraries in 1861 now reappears as a European necessity: allowing access so far limited to privileged librarian's hands and thus responding to the inquiries of readers. The most urgent measure is the protection of Prussian index cards against unauthorized consultation or withdrawal. "Only for official use / We produce / Order and security."[19] If bound catalogs bear the stigma of being transitory, this applies less to the card index—but only as long as it is guarded from the audience and used only by initiated librarians. "Catalogs were created by librarians merely for their own use, when they were the only mediators between users and books," as it was put as late as 1961 by leading librarians—nostalgically and not without regret.[20] The people who write the catalog have different access rights than the people in the reading room. Once it could be said laconically that "for understandable reasons, nobody is admitted to the royal and imperial court library without being accompanied by one of its employees."[21] The reader is a potential threat, a possible cause of disarray and disturbance. "I went on a little longer about needing a kind of timetable that would enable me to make connections among all kinds of ideas in every direction—at which point he turns uncannily polite and offers to take me into the catalog room and let me do my own searching, even though it's against the rules, because it's only for the use of the librarians."[22] What General Stumm von Bordwehr experiences in 1913 as a merciful exception had by then already become an everyday occurrence in Prussian libraries. "The public has now conquered the catalogs."[23] As soon as it becomes unavoidable to open the catalogs to the "user demanding information," the card index attracts drastic changes in the way it is used from the hitherto restricted access and becomes subject to wear and tear.[24]

Library administrators raise many doubts about readers' democratizing claim of free access to the catalog, which they nonetheless concede over time. There is at least a subliminal distrust toward intruders and their possible destruction of the established order. Not only are cards of inferior quality (compared to those manufactured by Library Bureau) in danger of rapidly wearing out—there is also the infinitely larger threat to the loose arrangement of the card index. Apart from thread or drawers, there are initially very few security measures for the paper slips. The possibility of misplacement causes considerable anxiety: "One must realize that the rod holding the index cards was not invented to prevent cards from falling out, although it may also do so every now and then." [*Paragraph. Eerie silence.*] "There is generally no suggested course of action in the rare event of a box of cards accidentally falling to the floor. The reason for a hole punched through the cards and the insertion of a rod is that the need to simply unscrew it suffices to prevent the careless withdrawal of cards: a principle that makes clever use of psychology."[25] The "step backward" to a slit system[26] as it was used in the rest of Germany was not carried out by Herbert Davidson—though he boasts (as already mentioned) that he was the one to introduce this first controlling bit[27]—since the slit slips of paper were "limp, delicate, and difficult to use when it comes to alphabetizing thousands of slips of paper quickly and efficiently."[28]

The objection to the slit that Ladewig appends to his praise of the Library Bureau (aiming at potential competing systems) in the end exhausts itself in suggestive turns of phrase, and again he praises the quality of the Boston product.[29] "These appealing, indestructible, movable boxes, always neat, space-saving, and pleasing to the eye even when stacked, are superior to any other sorting mechanism anyone might furnish."[30] Why then does the Royal Library in Berlin, despite Paul Ladewig's advocacy, not opt for Library Bureau systems and commission a carpenter to furnish card index boxes for the new facility?[31] Why was a veritable *safety barrier* erected to prevent unauthorized access by users and to secure library control, instead of establishing the punched card and rod?[32] One reason could be the absence of Library Bureau offices in Germany, or the insufficient proliferation of its products there. By 1894, the company had opened a subsidiary in a strategic location across from the British Museum, its first overseas branch, to cover Great Britain as well as the Continent with its standardization program.[33] However, in the German Empire advocates like Ladewig wait in vain for the opening of a branch to demonstrate and sell the appropriate devices.

What remains indisputable is the discursive failure of the public library movement and its proponent Paul Ladewig. His brisk pamphlets in support of public libraries make him Dewey's German spokesperson—but they are defeated by the sharp criticism of his opponent Walter Hofmann, who dominates public attention with his proposal for small "popular" libraries.[34] Ladewig finally concedes, "The vigor of the German public library movement, at that time often under the name 'standard library,' was significantly curbed."[35]

Though in Germany the demand for Library Bureau products seems to succumb to ongoing quarrels over suitable library supplies and catalog form, ideas about standardized and centrally printed slips of paper in uniform and extendable boxes gain hold in German offices. Eventually, European library administrations embrace the card index as a useful and efficient method for catalog management. Thus, the library tradition of data management and accounting that succeeds in business after 1890 and becomes a thriving industry in its own right finally returns to the library.

The Bridge Enters the Office: World Brain

The most difficult thing about collecting is discarding.
—Albert Köster

While the slogan "Free libraries for every soul" becomes a bone of contention between librarians' factions in Germany, the remaining points of Dewey's reform program are less controversial in the Old World. Perhaps the most important transmitter that spreads some of the American Library Association's arguments is in fact the Institut International de Bibliographie in Brussels, founded in 1895 by Paul Otlet and Henry La Fontaine, the latter having coined the phrase "memory of the world" in connection with the first international bibliographical conference.[36] "What the founders of the institute have in mind is, ideally, gathering book titles from all times and peoples in one central location, in two separate series, an alphabetical author's index and a subject index using the decimal system."[37] Since Dewey's decimal classification (an indispensable basis of the plan) is still regarded with skepticism in German libraries, the institute programmatically aims at its dissemination. Hence, the scheme includes "perfection and propagation of the Dewey decimal system [and] production of 'Bibliographia universalis,' i.e., a catalog comprising the world's entire book production arranged according to the decimal system on slips of paper by (a) copying and unifying the titles from *every* large library, (b) cataloging *every* new publication (books and articles)."[38] In addition to financial support by the Belgian crown, the capital for this project consists of 400,000 slips of paper in "boxes contrived after the American system"[39] that form the basis for a world catalog compiled by La Fontaine and Otlet.

The growth of the card index leaves little to be desired. By 1897, the universal register already contains about 1 million entries; by July 1, 1903, there are 6,269,750, and by 1914, more than 11 million slips of paper.[40] With an annual increase of approximately 500,000 entries, estimates see the project covering the "entire book production since the invention of the printing press," anticipating the end of this gigantic enterprise after another ten years.[41] It is no coincidence that the development of the *Bibliographia Universalis* is reminiscent of Konrad Gessner's efforts. As a method for the production of index cards "by cutting and pasting bibliographical aids," it undoubtedly harks back to this origin and bears a striking resemblance to its parameters even 350 years later—despite the improved copying procedures and printing methods, paper quality, and cutting devices that were developed in the meantime.[42] Following emphatic recommendations by the bibliographical institute in Brussels, the doctrine of basing data collections on semantic arrangements sorted with decimal figures is slowly adopted in business applications and, with some delay, also by the Prussian administration. "A number of companies in Germany (Brown, Boveri & Co., A.E.G. among others) have introduced [decimal classification] for their card indexes."[43] Electronics firms, the industrial avant-garde in the corporate age, follow the suggestion willingly, yet libraries remain hesitant. Finally, in 1911 a "preparatory committee" attempts to imitate the Brussels bibliographical system—copying individual concepts like *world brain* and the aesthetics of illustrations (see figures 7.1, 7.2)[44]—and its administrative imports from the United States.[45] This committee, later attracting considerable attention, was not aiming exclusively at librarians; it bears the telling name *The Bridge (Die Brücke)*, referring not to transatlantic or trans-European transfers, but to a worldwide transfer of information between centers of scientific activity. The goal is to establish connections between islands of knowledge and bridge the shallows in the sea of ignorance.[46] Although both institutes have similar aims, the committee sees itself less as the German-speaking branch of the bibliographical world catalog in Brussels than as an independent enterprise with its own original objective—for it aims not at collecting bibliographical information pertaining to old and new knowledge, but the atomized components of the very sources of intellectual work. Nonetheless, the intended field of operation—as in Brussels—includes the entire world, with The Bridge seeking to become its productive "brain."[47]

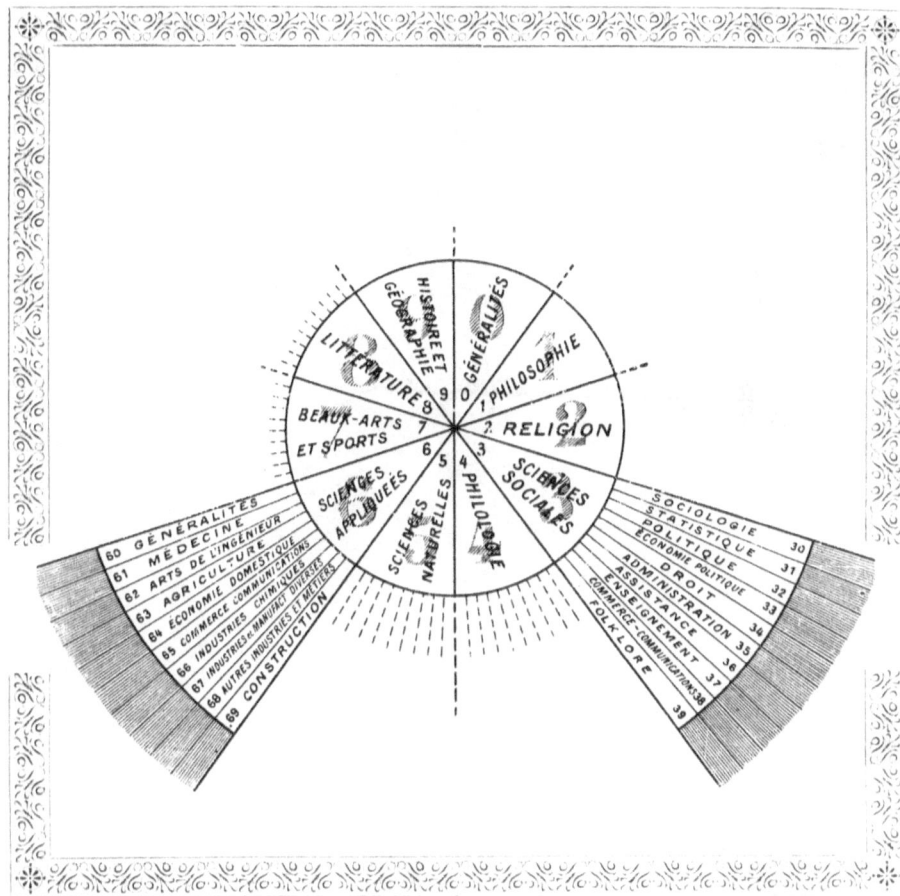

Figure 7.1
Dewey's scheme, displayed by the *Institut International de Bibliographie*. (From Institut International de Bibliographie 1914, p. 45.)

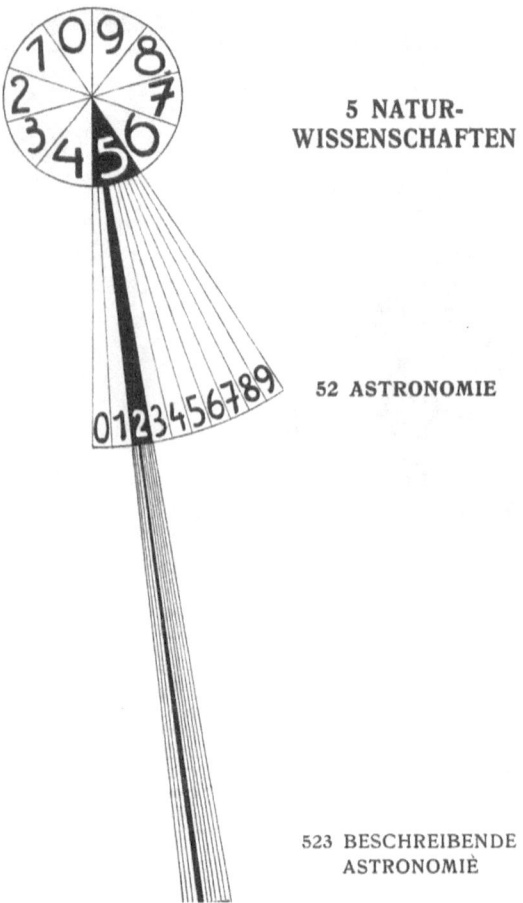

Figure 7.2
Dewey's scheme, displayed by *The Bridge*. (From Bührer and Saager 1912, p. 4.)

May 1911 sees the publication of *The Organization of Mental Labor by "The Bridge,"* by Karl Wilhelm Bührer and Adolf Saager. The authors' aim is unequivocal. "The principal purpose of our book was to win Ostwald for our cause"—namely, as a promising financier and mediator for knowledge workers.[48] Encouraged by having read Wilhelm Ostwald's book *Energetic Bases of Cultural Studies*, the journalist Saager succeeds in translating Bührer's ideas, despite communicative difficulties, into a joint text, and to establish contact with Ostwald.[49] In spring 1911, Saager sends printer's proofs to Ostwald's country house in Groß-Bothen.[50] Ostwald responds immediately; they meet and found The Bridge as an *International Institute for the Organization of Mental Labor*. The capital consists of 100,000 Reichsmark (a third of Ostwald's of Nobel Prize money) and a friendly donation by the industrialist Ernest Solvay.

The purpose of The Bridge is the avoidance of all superfluous energy spent in the entire area of pure and applied mental labors, and in connection with this, liberating the creative mind from the shackles of preliminary mechanical work.

These energy savings should be achieved by planned organization of mental labor, namely:

1. Through full exploitation of mental labor as done currently: by means of an international information exchange that should provide anyone with answers to any question;

2. Through suitable organization of the more organizable—that is, mechanical— processes of future mental labor: by means of an international organizing agency that shapes the forms of mental labor of all kinds so that every individual project fits *by itself* into the whole organism.[51]

Aside from the additional plans that Ostwald has for the institute—for instance, spreading the *energetic imperative*, as indicated in the above quotation, or promoting the metric system in the English-speaking parts of the world—one central point unites the otherwise very differently motivated initiators: namely, how to achieve these ideas.[52] They regard the written sheet of paper as culture's basic technical form. The *Monographic Principle* plans to take down every single thought faithfully in its ideal form, on paper, in standardized format.[53] "If mental labor is to be organized, one must begin with the organization of the memo."[54] Since bookbinders are the enemies of mobility, giving each free thought its own piece of paper to connect it with other building blocks of knowledge is essential. Thus, modularized world-knowledge is to be stored on monographic index cards at the "reference desk of reference desks,"[55] Schwindstrasse 30, in Munich. This will afford intellectuals the means to recombine monographic thoughts into new constellations.[56] However, before they can store any such knowledge, the slips of paper must all have the same size, the so-called *world format*—a precursor of the DIN and ISO formats of German industry standards, but a successor of the "mono format," the brainchild of a company that was bankrupt by 1911, and whose sales representative and chief administrator had been none other than Karl Wilhelm Bührer.[57]

The logic with which The Bridge proceeds to implement a world format is a clever syllogism. The Bridge can only become a reference center for world experts if world experts contribute, that is, share their results with The Bridge. The infrastructure must be established by specialists as well.[58] However, the most significant scholars of the world have other things to focus on than "solving organizational problems; [...] the execution is in every isolated case a matter of expertise, as it is everywhere, thus of the organizer and therefore again a matter for The Bridge."[59] Only The Bridge feels capable of this task, and thus it authorizes itself to take remedial action. Its organizational innovation movement pursues two strategic purposes: (1) An "archive that will introduce a comprehensive illustrated world encyclopedia on single sheets of uniform format." At first, world knowledge is to be put down by professionals on standardized slips of paper and kept in standard boxes in a world format.[60] (2) A "collection of addresses, containing the addresses of all living knowledge workers."[61] Furthermore, they aim to gather pointers to knowledge, which by virtue of their own addressing logic lead to new information.

Maybe the ads for Library Bureau and the organizational achievements they promise encourage a belief that with the help of a suitable apparatus and thanks to the considerable time savings that ensue, *everything can be stored.*[62] Operating an index card box tempts people to develop a euphoria of totality—and not only the The Bridge members. In 1897, the imperial librarian in south Berlin, Christlieb Gotthold Hottinger, announces a *Book Slip Catalog and a Bio-Icono-Bibliographical Collection,* nearly at the same time as Otlet's institute does so in Brussels, and presumably inspired by American index card printing.[63] The second edition of this announcement—parallel to that of The Bridge—in May 1911, once again presenting the idea of an archival storage *of the world*, especially in terms of knowledge work. Paper slips in the international format are to be indexed and sent to subscribers, not without also gradually including older printed works in Latin, German, English, and French. "Retroactively, anyone who has published written works or has been otherwise outstandingly active is taken into consideration."[64] To provide a location for the compilation of this "total card index" and to guarantee the future growth of the project, all information should enter the "paper slip headquarters (with typesetting, reproduction, clichés, galvano-plastics, printing press, bookbinding, warehouses, a bookstore, delivery, and editorial staff with a large reference library)."[65]

Unlike The Bridge, Hottinger's plan relies on decentralized and active cooperation, and Hottinger asks all "friends of intellectual life" to participate. "With every single slip of paper, its author etches his name into the human monument of intellectual endeavor."[66] The two projects, the Bio-Icono-Bibliographical Collection and The Bridge, are united by the field of operation they wish to occupy; they announce that their programs will cover no less than the entire world. If the world forms the totality of the scientific body of knowledge, *nothing* should be left over in the end. "Science cultivates a precision that tolerates no unaddressed remainders; in other words: science works with an eye to completeness." Besides the world, there is a second major category: "The totality of scientific thinking knows no restriction and [...] concerns the work itself, the object of the work, the space in which the object exists, and the time in which it exists."[67] Attempts to organize the world around 1900 are marked by a vision of completeness, though Hottinger limits the concept, while Bührer and Saager aggressively and clearly emphasize it. Now it is incumbent upon the aggregated index cards to achieve this completion.

All these attempts to create a "gigantic," "universal," or "complete" recording *of the world*, whose brief genealogy leads from Otlet/La Fontaine to the forgotten Hottinger and on to Ostwald/Bührer/Saager, are aimed explicitly *at a completeness* that in today's view seems bizarre. Why do ideas and data management projects like these not appear earlier? They certainly have not been fostered at other times by increasing amounts of material that challenge the limits of experience. On the contrary, voices warning of a flood of knowledge always abound, as they do around 1900: "Even under the best circumstances, a complete collection of material is always the exception. [...] Despite the utmost completeness of our bibliographical police registration offices, the utmost diligence, and the most careful mutual control of the work, single volumes are bound to escape the list."[68] This is a sober insight that need fear no contradiction from the perspective of contemporary common sense. However, the successful exception remains speculative and is valid only under the tightest restrictions—for instance, in an investigation of "silkworms in the district of Mörs."[69] Only "primitive, simple factual material can be complete once and for all, yet it is so only under such rare conditions that these have to count as exceptions. The secondary academic material is in fact never complete, for the trivial, unfortunately unquestionable reason that all our knowledge is piecework."[70] In the eyes of The Bridge, this argument reads less like an attempt at destructive criticism and more like an encouraging confirmation of the need for modular building blocks that amount to a complete system stored in a suitable memory bank. And it is presumably advertising for the memory procedures, already well established in the business world, that promise to save a lot of time by dint of a universal library technology. The contemporary library and administrative technology, the new office apparatuses encourage the pioneers of the world brain development in their bold projects. However, the card index, euphorically imagined as a mechanical memory machine, still obscures from view the robotic character of the machinery. Indeed, the input of records still requires a human being.[71] The initiators of these global data projects succumb to the belief that having the algorithm also means immediately having the results.

Thanks to diligent writing and reading operations, the registers experience a steady growth. "Our bibliographies for instance have bit by bit gained such a circumference that the full use of these complete lists is impossible,"[72] Richard M. Meyer objects angrily; his critique focuses soberly on the method, and he delivers the deadly argument of *selection*. "The easiest and most practical way to emphasize what matters is to omit everything that doesn't."[73] This is what Karl Wilhelm Bührer anticipates in 1907. "One completely forgets how to separate the important from the insignificant, and then makes the excuse that in science, everything is important."[74] Four years later, however, the general manager of the Mono-Gesellschaft states, "Organization will need to take pains to accomplish a complete precision that only a superficial man would describe as pedantic [...] The same attention must be bestowed impartially on every component; it must make equal room for a manufacturer's catalog, a poorhouse's chronicle, and the works of Homer or Kant."[75] Wilhelm Ostwald hesitates when reading about Bührer's view of science. He claims that Bührer hardly knows anything about science, yet fatally "[leaves] it at that."[76] When founding The Bridge and again later, Bührer insists on doing things his way, and Ostwald, in distant Saxony, has to give him free rein.[77] Ostwald is evidently distracted by various other activities—for instance, the Monistenbund, the antichurch movement, or the development of the *Glücksformel*, the formula for happiness.[78] Thus, some of Bührer's projects lack the required combination of pure theory and applied practice. With ideas like the "irrigation of the Sahara" (another totality project), the "cultivation of vast swamp areas," or the "reforestation of Greece and other civilized countries," seriously considered and declared as international cultural tasks, The Bridge moves closer to the *world archive for humor* than its founders might have intended.[79]

The function of completeness within the coordinates "labor" – "space" – "time" is fed expert knowledge to serve as a starting point for scholarly data processing. "Only a completeness of observation approximately guarantees that no lapse is been committed and that chance is eliminated."[80] Inspired by a system that omits marginalia—scientific thinking knows no secondary matters—The Bridge plans to prepare a world-answering machine. Its particular knowledge units await being absorbed into any possible connection—prompting scholarly paranoia: "Even the slightest fact is significant, even if one may personally be convinced that it is not, because it might lead at some point to future conclusions."[81] These stubborn assertions of the (academically untrained) former sales representative for advertising cards assume that a few mechanical tools can realize an all-embracing and world-encompassing collection of materials. Only the second strategy of The Bridge—namely, to compile references at the highest possible integrated level instead of collecting the material—points to Ostwald's yet unvoiced doubts about the attainability of completeness.[82] Nevertheless, Bührer's activities soon drown out Ostwald's skepticism. Bührer proves that he has already forgotten the difference between signifier and signified. Instead of storing the address of a music archive in The Bridge's card index (as metainformation of metainformation) or copying its catalogs (as a transfer of metainformation), he goes ahead and purchases the entire archive.[83] Tempted as he was by the idea of completeness, this was only the first, practical step to the realization of his plan.

The Bridge resembles Dewey's project in a quite unexpected way—its inevitable failure. The partners' financial plight intensifies when the expected donations by "very rich people" fail to materialize.[84] Meanwhile, Bührer leads the business inexorably into ruin. In June 1914, exactly three years after its foundation, bailiffs close the offices of The Bridge in Schwindstrasse 30. The proof of fundamental incompleteness will not be disclosed until 1931, when Kurt Gödel's proof of formally undecidable sentences is published.[85] Empiricism produces its joyless results sooner: the knowledge actually gathered in this ambitious world registration finally remains limited to a complete collection of postcards of the small town of Ansbach, tranferred by Bührer himself into the global format as proof of his strategy and of the efficacy of the memory bank. The last part of the case history: "A mere collecting frenzy takes the place of scientific work; a brutish greed for completeness."[86] At the trade show Office and Business in Munich in 1913, an astounded public witnesses Bührer's presentation of "mass-produced advertising stamps collected avidly by children, procured from everywhere and stuck on cards in the world format. First he carried this work out alone; then, however, owing to the abundance of the material he proceeded to enlist the staff of The Bridge."[87] After Ostwald settled the remaining business debts, he received the proceeds in postage-stamp-sized currency: "those countless glued-on advertising stamps. Soon afterward the war broke out."[88]

8 Paper Slip Economy

I had firmly grown together with my desk /
And firmly grown together with me was my desk.
—Heiner Müller, "Wolokolamsker Chaussee IV Kentauren"

Although The Bridge collapses on the eve of World War I, advertising stamps proving an insufficient foundation, Wilhelm Ostwald succeeds in drawing attention to his universal concepts of organization even after the failure of the project. However, in 1914, the global turn of events—that is, the outbreak of World War I—galvanizes the mental labor of nation-states into a different scientific and economic mode. In the German Reich, this is reflected in the test for index cards as a catalog technology when, in 1914, they replace the bound catalog of the Royal Library in Berlin to collect and order everything by objective keywords.[1] Even before the war, Ostwald's energetic economy[2] connects with the adoption after 1913 of scientific management as pioneered by Frederick Winslow Taylor, particularly by his German translator Rudolf Roesler.[3] Roesler emphasizes that both proposals "are applicable with the same right and with the same success to all areas of human activity."[4] The central point of their interaction and mutual reinforcement lies in the search for efficiency, for increased achievement. The energetic imperative suggests avoiding wasting energy (W), and moreover seeks to increase it. However, Taylor's studies disassemble larger movements punctiliously into smaller ones, analyzing the time needed so as to decrease it. The overlap of the two theories is not by accident the definition of *efficiency*; and after 1911 the *efficiency craze* in the United States becomes almost a pop phenomenon.[5]

In equation 8.1 (figure 8.1), the limit value of efficiency addresses dynamism in Taylor's sense, so that output (P) increases with reduced time (Δt)

$$\lim_{\Delta t \to 0} P = \frac{W}{\Delta t}$$

Figure 8.1
Equation 8.1: Energetics and Taylorism as a formula.

required for executing an activity—for instance, a greeting. Once theoretically firmly prescribed (i.e., standardized), this one clear instruction is enough to implement the rationalized action. "The simple process of the soldier's salute is defined as follows: right hand in the shortest time by the shortest distance to the headgear and then back to the trouser seam [...] This instruction normalizes the process of saluting and is the cause of exact realization everywhere and of a uniform appearance of the process."[6]

On the German side, the discussion of Taylorism and its (military) industrial application wanes, since transfer of knowledge and technology is limited during times of war. Yet at the end of the war, it starts up once more, and this time with lasting effects. Meanwhile, some successfully Taylorized companies in America produce noteworthy results that supposedly prove consequential for the war: Irene M. Witte, a student of Taylor's acolyte Frank B. Gilbreth and a defender of Taylorism during the Weimar Republic, cites Taylor's development of steel tool-making methods that allowed the American wartime economy to produce five times more ammunition than by conventional methods.[7] The German defeat results not only in broken-down industries formerly engaged in the wartime economy. Public administration also threatens to lapse into confusion. "Instead of simplifying office organization in times of great need, anarchy has taken hold of many a workspace."[8]

Many of the businesses affected directly by the Treaty of Versailles seize the opportunity for organizational reinvention. Equally, government administration has no choice but to tackle reorganization. "Let us not forget, however, that war was necessary to precipitate change."[9] Along with the arrival of the new paradigm *organization*, Wilhelm Ostwald strives for the integration of his doctrines of energy and universal procedures into a metascience. "The ordinal science or *mathetics* (not mathematics) is the most general of all sciences, and therefore in a certain sense the beginning of all wisdom."[10] To point to the roots of this wisdom, Ostwald illustrates his argument about the basic application of the energetic imperative with the evolution of the automobile. This concept serves like no other as a midwife for Henry Ford's production line idea, as well as for Taylor's concept of "assembly-line work," thus figuring as the starting point for all American management science and ergonomics.[11] Although Ostwald's later writings still refer to Taylor's work, the example demonstrates the interweaving of energy savings with ergonomics within the German postwar adoption of Taylorism. Artificial limbs for the disabled and aptitude tests for truck drivers (introduced by Georg Schlesinger) along the lines of *psychotechnics* point to savings of time and energy by virtue of closely analyzed and rationalized motion studies.[12] A growing number of institutions such as railroad companies,[13] water works, gas and electric utilities, and ultimately the Prussian government—with its compilation of penal registers using Bertillon's method—praise the new organizational achievements in terms of the "machine principle of time saving"[14] and *rationalization*.[15]

However, in this final chapter we will not explore the complex development of scientific office management.[16] Rather, we will focus on a product catalyzed by office management and placed at the center of attention: the card index as a handy and surprisingly useful aggregation of paper slips. How does the organizational discourse succeed in giving a traditionally established principle the surprising appeal of irresistible novelty, promising an unprecedented increase in administrative efficiency? How is the principle of indexing liberated from its prehistory and made new to satisfy the increasing market for office machines and its moneyed clients?

System / Organization

The basis of organization is to ensure that all work continues to flow on its own.
—Karl Scheffler, *Neuzeitliche Büroorganisation*

As we have seen in previous chapters, technology transfer between the industrialized continents, which up to the year 1890 had occurred mainly from east to west, begins to take place the other way around in the shape of a lasting scientific management and ergonomics discourse.[17] Concepts such as *efficiency, rationalization, standardization* arrive under the imported paradigm of *organization*, and the prevalent *principle of the shortest time* asserts itself in terms of both voluntary and unavoidable reforms of European institutions.

The year 1900 sees the founding of *System: The Magazine of Business* in Chicago, and by 1910 the journal has dismissed contemporary office solutions in favor of the loose-leaf binder,[18] advocating "card index systems" and praising these devices as an immense achievement compared with conventional filing systems.[19] In 1902, a British edition begins publication, while the German edition, published in Berlin under the name *System: Zeitschrift für moderne Geschäfts- und Betriebskunde in Handel und Industrie* starting in 1908 and primarily focusing on typewriters, fails after only two years. Nonetheless, the journals do have an effect. Recalling the dissemination of ideas by the Institut International de Bibliographie and The Bridge, the American "modern office organization" procedures manage to get people on the other side of the Atlantic talking about typewriters, carbon paper, card indexes, and loose-leaf books, and these office aids are readily adopted and optimized by European companies.[20] "First they were looked upon rather critically here, then they were installed in German offices and thoroughly improved. When their usefulness was proven, they found entrance into home offices. According to this development, trading ventures were the first to adopt these new aids."[21] At the International Exhibit for the Book Trade and Graphic Arts (BUGRA) in Leipzig in 1914, the "abundance of new office techniques and aids attracted a lot of attention."[22] Of the 133 exhibitors in the hall for stationery, writing tools, and paints, only three (Organisations=GmbH, Albert Osterwald, and Autoclip GmbH) devote themselves to the production of card index registers and accessories.[23] These putative innovations as well as loose-sheet account books attract more attention than typewriters, postcards, erasers, firecrackers, flowerpot covers, and so on. The hardly objective report by Friedrich Soennecken of Bonn emphasizes the special position of the office industry with its various pivotal achievements. Not surprisingly, he mentions the exceedingly successful letter file from his company's own line of products, followed by his competitor's: "Germany also made headway with card indexes and loose-leaf accounting systems originally developed abroad."[24] In an equally prominent medium, the *Official Exhibition Guide*, Albert Osterwald as one of three index card producers personally summarizes his stroll through the office supply hall. He sees above all "index card registers for inventory control, accounts receivable, prospective customers, etc., as well as modern index card accounting systems, like Osterwald's automatic accounting that eliminates errors."[25] This self-referential report raises the suspicion that it is merely an advertising scheme to stimulate demand for these yet-unused office tools. What Max Frank describes as an abundance of office-technical innovations at the BUGRA seems to apply less to card indexes and accounting systems than suggested.

The advent of World War I paralyzes any favorable adoption of new office technologies.[26] Card fabrication serviced by a few young manufacturing companies declines temporarily. A wartime economy builds up and increasingly takes over production capacities. Only after 1918 does a real demand-and-supply loop emerge, giving rise to a number of companies providing office supplies. The restructuring of many enterprises, enforced by expropriations or the chaotic aftermath of war-crippled management ("administrative knowledge was suffocated"[27]), produces a need for new management concepts. It is the task of advertising in technical periodicals to stimulate and meet the need for the new office products. Relaunched magazines like *System: Zeitschrift für Organisation und moderne Betriebsführung* or *Organisation—Betrieb—Büro* are joined by new publications, including *Zeitschrift für Organisation* (1926) and *Wirtschaftlichkeit* (1926), neither of which fail to honor and praise their pet child, the card index.[28]

Universal / Card / Machine

With the incipient reorganization of the German Reich after World War I comes not only a general paper flood, particularly in the form of new decrees for "suffocated" businesses; there is also inflationary bank note circulation.[29] To regulate and manage this tidal wave of paper, the organizational literature that proliferates almost to the same degree as money concentrates on the card index as a powerful achievement.[30] Its advantages range from simple savings in time and money to virtually universal application, its negative counterpart being the book, a seemingly outdated concept. "The card index overcomes the book. Its proper characteristic is vertical order."[31] With this upright liberation from the configuration of bound sheets, defenders of the card index begin to celebrate the advantages of loosely arranged and cardboard-reinforced slips. Some of these advantages are listed here with all the brevity recommended in management theory.

"Card indexes are books broken up into their components."[32] Following the card index's victory over the bound book, advocates emphasize its functional superiority of mobility. Independent entries allow unlimited insertions, constant extensions, and reconfigurations, while continuously keeping data up to date.[33] The correct choice for one's company among the variety of systems offered can be made only on the basis of a distinction between writing and reading.[34] The first type, a *reading index*, is good for accumulating information, stored every now and then as *read-only memory*. Just as the book used to be considered a depository of knowledge, so the card index replaces the book by means of a more adaptable and "mobile memory."[35] An intermediate position is occupied by the second type, the writing card index, because information is stored briefly and disappears just as quickly. Its storage features remain arbitrary and allow for a *random-access memory*—however, it is precisely addressed. With the aid of sorting criteria such as the strict alphabetical ordering of keywords, customer names, and the like, access is considerably accelerated: each card is immediately at hand and, not least, always offers suitable opportunities for checking.[36]

With the arrangement of modularized slips of paper, access to the slip boxes undergoes a change as well. Consultation of the card index occurs from another perspective: the book may not force sequential access, but it certainly suggests that path (though it does permit browsing); by contrast, the card index requires ipso facto a different perspective. Extracts from its data occur by recombining the modularized entries, permitting an easy overview—and it is exactly the absence of this overview that is posed by conservative bookkeepers as a counterargument.[37] Yet the variations afforded by level card index or view card index technologies eliminate this defect.[38]

The frequently emphasized benefits of the allegedly innovative apparatus arise—unsurprisingly—from saved working hours. However, if scientific management typically measures every unit of time, psychotechnical investigations of card index systems fail to refer to the inevitable time required for data entry and training that need to take place until a routine is established for this "sensitive indicator."[39] Instead, there are long lists of the characteristics required of the "card index leaders," that is, the people responsible for the card indexes, whom we would call system administrators today.[40] "There is no tool that is as worthless in the hands of the clumsy user as the card index."[41] The card index appears as an intricate and demanding machine, for whose use a skillful and responsible organizer is vital. It is incumbent upon him to instruct other users, with the help of precisely laid out instructions. "Mechanical workers become masters of their work if the organizer does the thinking for them."[42] A decisive incentive for the complex deployment of card indexes in business is their ability to grant several users simultaneous access to independent commercial proceedings.[43] This is multitasking in a multiuser system: an order is no longer limited to one authorized bookkeeper successively entering transactions, but exists asynchronously alongside numerous other processes, handled by other workers—as promoted by the Taylorist concept of the division of labor.[44]

Meanwhile, the press unanimously praises the financial and efficiency benefits of the card index over the laborious noting down of proceedings in heavy bound tomes. It is nevertheless surprising how quickly the praised administrative technology obliterates its own history. The marketing literature finds no tradition for the card index worth referring to; it mentions no historic instances of use. As there is no authoritative reference to history, only one other strategy remains for proclaiming the product as an unconditional innovation, as a compelling break with tradition—in short, as *modern*.[45]

With their propaganda, the first companies that attempt to conquer the market for organization make it clear to office workers that their current production capacity lies far below an optimum level. Yet to achieve the maximum, new procedures, aided by new machines, are needed. Anywhere this technology is already in use, numerous improvements, according to the logic of "updating," seek to save even more time. However, one obstacle stands in the way of an ideal time savings: for fear of being replaced themselves, unwilling bureaucrats defend themselves against subversion by machines that threaten to take over their work. "Was that my desk or was it me / Who whispered it to you, old Prussian / [...] / My desk and I, who belongs to whom / The desk is state-owned What am I / Below a desk, above another person / No longer a person but a human machine / A furniture person, or a human piece of furniture."[46] (See figure 8.2.) Reinforced propaganda—"Does the office machine devour souls?"[47]—counters the attacks against progress: "Anyone opposing calculating or other time-saving machines in the office because they are an 'innovation'—old methods and entrenched laziness naturally regard 'innovations' as hostile!—will nevertheless sooner or later be forced to replace manpower

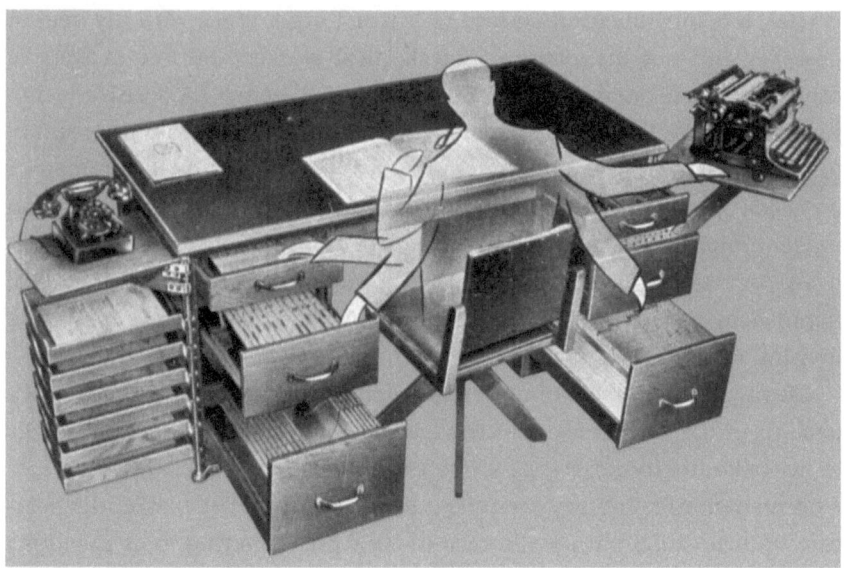

Figure 8.2
Furniture person. (From Schmidt 1939, cover page)

to a great extent by machines in the office, as rationalizing always means mechanization."[48] Favorable balance sheets give rise to more propaganda, that is, a sales strategy that expounds the indexing principle as an innovation, although—as this study attempts to show—the underlying procedures of indexing in fact have a long, undisclosed tradition. Only new concepts, as the advertising maxim suggests, are marketable.

The market guarantees enormous long-term profits, for there are still areas of business that have not been accessed. Administrative offices, manufacturers, and industry offer considerable potential for the deployment of card indexes so as to benefit both customers and salespeople:

> The card index is needed especially in current accounts and accounts receivable, because it guarantees controls. Furthermore, the card index is useful to record customers, to track orders, sales representatives, revenues, sales, and offers, for purchase and warehousing, for buyer statistics, for sales statistics, for storage management and dispatch control, for payroll accounting and benefit management, as a personnel register, as an arrangement card index, for loan control, check control, as a signature book, for credit accounting, for account deposits, for purchases and sales of stock, as referral control, as an expiration control for option deals and deposits, as a tax index, to track deaths and patients' files, as an insertion card index, a

stereotype card index and form card index, as a price-list card index, as a register of all kinds.[49]

However, before the card index legitimately conquers each and every business area and evolves into a universal (office) machine, a law is quashed, an event that will be discussed here in some detail. The law in question is not altered overnight. Rather, its effect is slowly vitiated by the practice of keeping ledgers, the basis of bookkeeping, on loose sheets of paper.

Invalidation
INVALID as the authorities say.
—Heiner Müller, *Der Auftrag*

According to §43 of the new code of commercial law that comes into effect on May 10, 1897, German businessmen are obliged to keep their books in a living language and with an eye to form: "Books are required to be bound and numbered sheet by sheet or page by page with sequential numerals."[50] Until 1897, §32 of the trade law prescribed the mandatory use of bound books; with the new version the "must" mandate becomes a "should" rule. The motivation for this seemingly tiny alteration that, according to commentators on this law, refers not to the ledger but to auxiliary books, is the institutional and transatlantic use of the card index as a key instrument for efficient accounting and documentation of commercial proceedings.[51] The card indexes sold by Library Bureau prompt the bookkeeping departments of nearly all American firms to convert to loose-leaf or card ledgers within a decade.[52]

The troublemaker breaking with the long tradition of folios as a basis for German bookkeeping is the "permanent accounting book" made up of loose sheets (see figure 8.3). The name hints at the history of the card index (permanence) and camouflages the new format's strategy by attributing the supposedly redundant word "bound" to the name of its opposite. From now on, the card index competes for profitable application in trade with the immobile, "large, thick, hard, and unwieldy"[53] ledger, paradoxically "book"-keeping on loose, uniform, and interchangeable sheets.

Advantages of the card index praised around the turn of the century include lower costs, a better overview, and greater security. Transfer mistakes become obsolete thanks to the merging of records that were entered onto cards.[54]

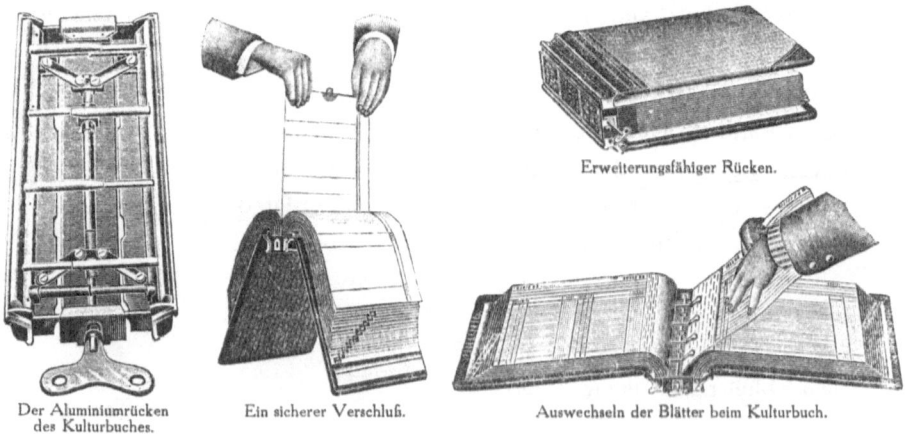

Figure 8.3
Permanent accounting books. (From *System: Zeitschrift für Geschäfts- und Betriebskunde* 6:133 (1909).)

Without going into detail here, let us note that businessmen and contributors to professional organs for office organization are the proponents in the quarrel revolving around the interpretations of §43.[55] Opponents temporarily include the Prussian chamber of commerce, as well as the imperial ministers of finance, who reject this innovative form of conducting business. Eventually, the dispute concentrates on the essential media-historical question, "What is a book?"[56] In the administrative parlance of 1908, the sober answer is, "In the expert view, and above all according to the latest court authority, a bound book needs to be equipped with pages that are sequentially paginated, and it must be bound so that no one can remove a sheet from the book without visible damage to the book and the sequence of page numbers."[57] In these terms, the permanent accounting book fits the definition. Novel implementations emerge from this definition, though, such as book covers held together by screws that allow a book to be rebound (see figure 8.3). Within twenty-five years, more and more firms use loose-leaf books to document their business proceedings—first in fields that are not immediately linked to §43, such as culture: "It is evident to any expert on currently available systems that the permanent accounting book is indicative of economic and cultural progress. [...] The 'culture' book offered by the company G. H. Rehfeld & Son, Dresden, has the benefit of such a strong mechanism that, quite clearly, one can lift the entire ten-kilogram volume by a single page."[58] Eventually the more advantageous procedure ousts the conventional form by circumventing legal regulations. Complemented by copying procedures that transfer card entries into ledgers and thereby prevent errors, the permanent accounting book comes to dominate bookkeeping, subverting the law by means of widespread practical application.[59] "The chamber has agreed to the admissibility of loose-leaf bookkeeping under the condition of the preservation of the principle of order and completeness," reads a commentary in perfect *Nominalstil* in the journal *Rationelle Betriebsführung*.[60] By 1929, it becomes clear that the use of permanent accounting books and consequently of index card registers will have no legal consequences.

Already by the following year, *Büroorganisation* trumpets in anticipation of a possible storage area for discarded accounting media: "Bound accounting books for double entry bookkeeping will be relegated to book museums and accounting museums. They shall be replaced by loose sheets of paper in folders, index cards and index card books, etc."[61] The tacitly subverted law grants a fragmented, distributed, and highly adaptable space to the basic accounting technique, whereby the psychotechnically optimized organization of the universal card index machine can rule. "Today no field of application can be imagined for which the use of a card index would not be of assistance."[62] Thus, although the German term denotes sheer disarray, business literally and comprehensively becomes a *Zettelwirtschaft*, a "paper slip economy."

The War of the Cards: Copyrighting the "Card Index"™

While the auspicious application of the card index at the beginning of the century is met with voluble recommendations, and there is much talk of accumulating index cards, the word *Kartei* (German for "card index") appears infrequently after 1904 in office organization texts. Instead, evasive names like *Kartothek*, *Kartenregister* ("card register"), *Kartenanlage* ("card arrangement"), and *Zettelei* emerge, all referring to the same apparatus.[63] The sudden disappearance can be blamed on the Prussian patent office, which grants Wilhelm H. Bach of Hinz Büromaschinen GmbH in Berlin-Mariendorf trademark protection on the word *Kartei*™. Yet after Hinz GmbH renounces exclusive use in 1909 so as to keep the phrase part of the common German language, a small company succeeds in securing the rights for its sole use. Unlike Hinz, Sponholz-Duca GmbH of

Berlin-Charlottenburg wages a forceful struggle and demands royalties for every illegitimate use of the term.[64] Especially Walter Porstmann, standardization theorist, figurehead of the card index movement, and author of a treatise on *Karteikunde* (the science of card indexes), receives countless warnings and reminders, and he is threatened with "severe punishment or a prison sentence of up to six months" if he refuses to stop using the term.[65] The "card index doctor"[66] Porstmann, a dogged advocate of German-sounding words, refers to the genealogy of the concept, which—although it was supposedly coined in 1904 by W. H. Bach—quickly became the common property of all German speakers.[67]

This genealogy is clearly insufficient. Apart from the fact that Konrad Gessner already mentions *libris chartaceis* and circumscribes the concept *in nuce*, Porstmann needn't have gone so far back.[68] As a peerless card index apologist, he restricts himself to a highly compressed economic history,[69] losing sight of the discursive origins. For instance, Porstmann could have found the Vienna court library act establishing the catalog of 1848, which says, "The whole mass of paper slips is divided into two card indexes (*Kartheien*)."[70] Numerous technical periodicals identify with Porstmann, the main target of the attacks, and join together, calling for civil disobedience and opposition against the copyright law.[71] "Use the free German word *Kartei* and regard anyone who tries to prevent you as a sinner against our precious German language, which is the people's property!"[72] By 1929, the German word *Kartei* for "card index" can again be used by anyone without fear of prosecution. "The years-long struggle a single company waged for sole possession of the word *Kartei* is over for good. How much money this pointless struggle must have cost! By decision of the imperial patent office, complaint department I of April 5, 1929 (reference S 11 344/32 W.Z.), the word *Kartei* has been expunged as a trademark and declared a free sign."[73] Thus, a different form of card index *reentry* ends in failure.

What remains remarkable about this protracted quarrel is the alternative nomenclature that produces peculiar coinages like *Zettelei* or even grandiloquent German compound words like *Sichtkartenregistereinrichtung* ("visual index card registry device").[74] Accordingly, the outrage against withdrawing the term from common usage is considerable: "Everyone regards it as an ordinary German word, just like *Bücherei, Auskunftei, Schriftei, Blattei*."[75] The alternative allegedly developed from *Kartentheke* is unacceptable to

Porstmann, the linguistic hygienist, owing to its "dented form *Kartothek*, which smells somewhat foreign."[76] For him, the term *Kartei* is a legitimate German coinage, though he loses sight of its etymological roots, which can hardly be called German. Rather, *Kartei* stems from the Greek χάρται, meaning "papyrus shrub," from which the word "cartel" derives as well. That the term *Kartei* introduced by the presumed wordsmith Wilhelm H. Bach (who was Porstmann's employer) was modeled on *Bücher-ei* ("(small) library") does not tempt the latter to consider that the word *Kartothek* might have been formed by analogy to *Bibliothek* ("(large) library").[77] This is supported not only by the substitution, which we shall now turn to, as the last episode in the (terminological) establishment of the card index.

Depiction / Decision

Now everything indicates that the book in this traditional form is nearing its end.
—Walter Benjamin, "One-Way Street"

Porstmann's foreshortened etymology raises an unobtrusive suggestion that does not fall prey to the fantasy of completion that caused the failure of enterprises like The Bridge. Just as resorting to the apparently *German* root of the word *Kart-ei* was expected to increase the attractiveness of the apparatus, so the smaller dataset a *Bücher-ei* holds in comparison to a large library or *Bibliothek* guarantees easier handling of the apparatus. "The resulting *Kartei* is nothing unusual, since we have in *Bücherei*, as an aggregation of books, already a suitable model for *Kartei*, as an aggregation of cards."[78] Why then is *Bibliothek* not the model for an arrangement called *Kartothek*? Porstmann operates with a large/small differentiation that marks the difference between *Bücherei* and *Bibliothek* in German usage around 1920. While the latter is a barely accessible, uncontrollable monster associated with a major institution, like the Royal Prussian State Library or large university libraries, the movements fostered by Ladewig and Hofmann, for example, emphasize the opposite: the small and accessible library, the *Bücherei*.[79] With its preselected holdings, these smaller libraries aim less at storing the entirety of printed knowledge than at distributing selected knowledge and making it accessible to everyone. Thus, Porstmann discourages the use of the word *Kartothek*, with its cosmopolitan connotations: "It has a foreign ring to it, which is why people enjoy using it."[80]

It is essential to follow the relationship between the terms *Bücherei* and *Kartei*, in order to render the installation of a card index practicable. For the determining factor for the analogy is the initial concept and the number of references a card index can manage. A decentralized, accessible *Bücherei* seems conceptually better suited than a huge, central, traditional, and infinitely complex *Bibliothek*, whose endless flood of material has overpowered various card- indexing aspirations in the preceding centuries.[81] Porstmann's tactical support of the word *Kartei* aimed at securing the card index an omnipresent place in the office world, rendering it as widespread and common as a small public library. Porstmann suggests the feasibility of a card index system in parallel to the budding *Bücherei* movement, thus abandoning the traditional connotations of differentiation. Hence, it is barely surprising that *Karteikunde*, his manual for the efficient handling of a card index, contains no hints whatsoever as to a possible genealogy of the card index from library technology, let alone any technological transfer. The card index appears in the office organization discourse without the shackles of its library tradition, doomed to youthfulness and novelty.

Yet it was not only Porstmann's opting for the term "Kartei" that indirectly brought the card index apparatus into widespread use. It is also a functional regulation that speeds up a far-reaching media change, a basic rearrangement of values. It translates words formerly linked to ledgers into numbers that, removed from strict sequence, become a movable basis for calculations. The book is dismissed as a medium in favor of a memory arrangement made up of movable paper slips that serve as a central point of operation for the new science of order and organization.[82] "Every new system assumes a transfer of accounting documents into figures, just as the shift of accounting concepts into numbers has generally gained importance in modern bookkeeping."[83] The coding of standardized concepts into numerical keys indicates how the card index thus catalyzes a trend that culminates in the universal machine, the "card index machine" of Hollerith or Powers.[84] However, before these take up their declared "final goal [...] of reaching the final result faster, fastest," let us look at some of the consequences of the intimated media change from book to card index.[85]

Already in 1914, Wilhelm Ostwald recognizes that "in the office and in the factory, the transition from the book to the card index has already taken place," while in 1927, Rudolf Päpke, writing the first editorial for the magazine *Büro-Organisation*, looks back on decades of struggles.[86] By 1922, the war of modern writing systems is by no means decided in favor of the card index. On the contrary, despite its propaganda efforts, the card index only haltingly manages to put its own arguments into practice. Victor Vogt, founder and owner of Fortschritt GmbH in Freiburg, analyzes the problems associated with the book, which he believes ought to be displaced by the card index:

The full efficiency of the book is realized when the *linear* arrangement of material is involved. The instances are numerous and richly varied. This onefold diversity mastered by the book can assume many aspects: places, human beings, money, material, amounts, dimensions, etc. Since the relatively simple thinking that prevailed in the last millennium was generally limited to one dimension, it is only too natural that the book was dominant in this period. Recording is always linear. Chronicles are simply extended: the recording of chronologically successive data is best carried out in a book, where individual signs and lines are assigned successively over the course of time. *Reading, thinking, speech* are primarily one-dimensional.[87]

The *book* is consciously presented as an enemy here. The long practice of arranging facts, according to the bent of Vogt's media-materialist criticism, is the inverse of what should now be recognized as contemporary: allowing the original material, fragmented on paper segments, to grow in a sequence with the aid of an ordering schema. A book cannot ever provide loose and insertion-friendly arrangements in alphabetical order; glue holds together those things that, according to the dictates of time, belong together. The ambivalences and complexities of cultural history, its breaks and anachronisms, shrink into the materiality of an endless paper ribbon, the basis of the paper machine:

The contents of a bound typeset work are to be connected line by line. If we cut up every single page and glue one line to the next as we go along, what becomes of the page is a long and very narrow strip of paper. However, the pages are to be *completely* disbanded into these strips, and then connected page after page. A novel turns into a strip of paper of a kilometer's length and two to three millimeters' width: it becomes linear, as its width becomes irrelevant.[88]

The inadequacy of the book having thus been declared, salvation can be found only in the card index, whose *multidimensional* representation remedies the shortcomings of the book. It exploits the disadvantages of the book, only to finally unite the peculiarities of many books.[89]

This relatively early diagnosis of a crisis of linearity, with its unexpected anticipation of theoretical arguments resembling the hypertext debate of the last few years, inspires card index doctor Porstmann to formulate in *Karteikunde* a modest "philosophy of the writing surface."[90] On this level, instigated (*angezettelt*) by the weaving of a texture, the development expands into a spatial network, as it were, of commercial fabric.[91] Recalling the associative technology that connects the current-account card index with a suitable spot in the customer address index via a cross-reference, for instance, and thus breaks through the two-dimensional surface of the card index, a textual theorist and master of the fragment concludes in his own unconventional logic, "The card index marks the conquest of three-dimensional writing."[92]

By virtue of higher complexity, the requirements of managing economic affairs around 1920 demand a mobile memory arrangement. "Linear accounting lines through the complex will no longer do."[93] Thus, Vogt is forced to admit, "The book has reached the limit of its efficiency."[94] Yet in 1927, Albert Predeek, as a librarian apparently blind to this threat, still optimistically believes the enemy is somewhere else: "It will be a good while until the end of the millennial empire of the book, when instead of the 'handbook' in every field of science a huge card index of loose-leaf treatises will be arbitrarily arranged; though we can already see in the extraordinary increase of magazines a result of the abandonment of the book."[95]

Only Walter Benjamin logically sketches the scenario of an obsolete object caught between contemporary business and learned memory systems: "And today, the book is already, as the present mode of scholarly production demonstrates, an outdated mediation between two different filing systems. For everything that matters is to be found in the card box of the researcher who wrote it, and the scholar studying it assimilates it into his own card index."[96] From here it seems to be only one more small step for the detour via the scholar, the inserted learned subject, to be overcome, so that card indexes can hook up to card indexes and establish data streams whose transfer is less susceptible to mistakes. Subjects as well as (accounting) books are always potential triggers for errors, because the word serves them better than the protocol and control of numbers. "Transfers are sources of error. Moreover, transfers that occur periodically rather than necessarily cause a paralyzing lag, never being finished, and turn bookkeeping into what it shouldn't be: the writing of history."[97]

The installation of the card index in offices and stores succeeds with the aid of advertising strategies that recommend its products with no prehistory or genealogy attached, as ever more advanced innovative achievements. At the same time, the history of the card index disappears, along with the book, from accounting. The defoliation of the book frees the businessman, ignorant of library technology, from any knowledge about the development of his preferred tool. What it records are no longer characters of a living language—as §43 HGB 1 required—but merely figures without histories. In its business application, the card index may no longer describe the past; it need only calculate figures (and at most, store them). It is a pure calculating machine. From this exclusive reorientation toward numbers, the development of automated mechanical accounting ensues almost unnoticed: initially, punch cards are employed for bookkeeping in large companies; then they become a medium of universal memory, read electrically, and take up a systemic place under the invalid term "bookkeeping."

Hence, the attempt to outline the evolutionary development of the card index—from its supposed origins in sixteenth century Zurich all the way to the desks and card index troughs in offices around 1930—must end here, at the threshold of electrified calculability on a binary basis, on the eve of the universal discrete machine. Only the usually tacit use of card indexes beyond accounting, "for all independent professionals such as doctors, authors, etc.," prevents us from being unable to tell these stories. At this very concrete historical point in time, this collection of data-processing episodes on preelectric slips of paper also comes to an end, but not without a brief summary and a short preview of fully electronic connections.

Summary: Order / Cleanup

At the beginning of this study, I promised to describe a history of the card index as a foundation of the universal discrete machine. The starting point was the question, why all of a sudden is one object supposed to do justice to every data-processing requirement? In addition, I have attempted to sound out how the logical basic elements of a paper machine coincide with the development of slips of paper arranged in boxes. I have tried to point out how a theoretically endless, partitioned tape comprising uniform pieces of paper (representing books) develops by means of a precisely specified procedure (cataloguing instructions) and precisely adjusted read/write heads (library servants and scriptors) into an export product that opens a completely new market. The fact that writers and readers in this study appear only (and remain) as controlled subjects must rightly count as an inadequacy, yet this can perhaps be ascribed to the preelectric era: once they are electrified, writers and readers can be liberated and transformed into artificial intelligence.

In the second step, we saw the basic transfer of the procedure as demonstrated in the paper machine, a transfer that turned out to be equally profitable for accounting. I explored the dissemination of a "paper slip economy" in libraries as well as in businesses across two continents, indicating that the card index's claim to superiority in the business age relied on time-saving propaganda. The promised optimization was linked to a simultaneous rationalization movement as part of the paradigm of its time. Defenders of the card index believed in replacing the book as the vehicle of civilization, hoping for a media shift via hybrid composites so as to free traditional storage methods from the constraints of linearity.

If the main thesis of this study was the description of an information technology transfer, a second concern was examining the success of these transfers. Developments have shown that a history of the card index cannot avoid accounting for a preponderance of failed transfer attempts. The list of failures is seemingly endless: Gessner's catalog compilation brings him fame, but not the income he desires. Joachim Jungius's paper slip collection records every new fragment of knowledge, yet produces no new writing. The Josephinian card index fails to produce a bound catalog and thus fails to replace and eliminate itself. The French national bibliography is foiled by scant data return. William Croswell's long-delayed work, and his unpremeditated development of a paper slip economy from a mindset of slothfulness, lay the basis for an American storage technology, but this pioneering achievement goes unrecognized. None of the businesses under Melvil Dewey's own management—except for his retirement home project—are crowned with success. Paul Ladewig's attempt to transfer the public library movement and Library Bureau devices into German library management bears no fruits. The Bridge requires no further mention. Finally, even this study is presented as a bound book instead of a card index.

The transfer succeeds only in the margins of intention, in accidental deviations from deliberate applications. Thus, the balance sheets of the

Library Bureau owe their late boom to an accountant's arbitrary observation, helping to sell card indexes to businesses as well as libraries. And again, this market is discovered by Herbert Davidson more or less by accident, only to be explored all the more fervently in due course.

In the meantime, card indexes have disappeared from nearly every desk (and from nearly every individual memory) and have been replaced by uniform gray boxes, true universal machines.[98] The story to be told was also a data-processing history of progress, thus returning after various rearrangements to its starting point, an advertisement for an increase in mobility. Eighty years later, the card index writing this book requested the pamphlet "Bewegliche Notizen" ("Mobile Notes"), and after having received it in slightly altered form (see figure 8.4), added it to its inventory.

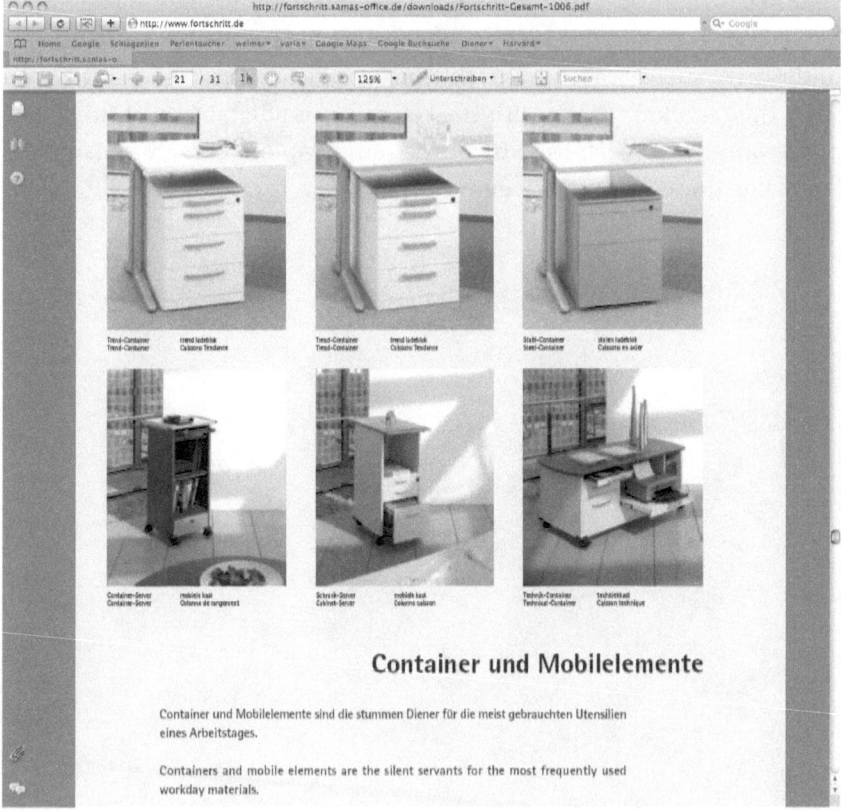

Figure 8.4
Progress. (From http://www.fortschritt.de.)

My initial intention was to conclude this history of data processing with the onset of electric procedures. The card index machine progresses inexorably after 1930 toward substituting its data carriers with punch cards and electric procedures to keep pace with new industrial standards.[99] Nonetheless, an ever-thinning line continues in the wake of progress up to the present day that preserves the continuity of card index technology. It is left to the residues of a recent office reform to unfold its effects in operational areas this side of electronic systems—for instance, in the preservation of monuments in the German Rhineland, or in the compilation of a "Card Index of Jobless Qualified Germanists and Germanists Threatened by Unemployment."[100]

Major libraries, like those in Vienna and Berlin, were recently seized by the winds of electronic progress.[101] Their voluminous card catalogs have vanished, making way for space-saving electrified terminals. Catalog rooms remain dirt-free, and deserted. In 1920, when library shelves are still state-of-the-art, a single voice heralds the oncoming electrified age that will not spare the libraries: "Along with stocktaking, it is advisable to clean books, repositories, and halls, be it by dusting and wiping, or by means of the lately much-used electric vacuum cleaner."[102]

Afterword to the English Edition

Progress knows no borders. Beginning with vacuum cleaners and telephones, electric and electronic technologies have invaded the library and have transformed information technology as well as the tools of rapid knowledge production. Be it in the form of catalogs that turn into OPACs, be it computerized data management, things no longer work without digital impulses.

And yet they remain the same—for instance, when one attempts to write the history of a reference technology and its many transformations. Its episodes and building blocks wander from books onto single slips of paper, are accumulated in digital or paper form, only to congeal, rearranged and formatted, into book form again—a laborious procedure. Nonetheless, an essential quality of the card index is precisely the efficiency of this system. The economy of signs allows us to feed on them over and over again. Though the care of the card index demands its own time, it can be regained in the end. "My card index costs me more time than the writing of books themselves," Bielefeld systems theorist Niklas Luhmann admitted, as perhaps the last analog card index theorist.[1] One might be surprised at first, given a publication list as prolific as his, unless one acknowledges the card index as the furnace in which the texts are forged.

"In principle, it is an infinity," as Luhmann described his communication partner, and thus it makes perfect sense that numerous entries and stories linked to them had to be left out of this revised and updated history of the cultural technology of quotation management. Owing to the strictures of book publishing, they must remain in my card index. For instance, the further development from punch cards and floppy disks to the USB memory stick extends from around 1930 to today as the gradual electrification of data processing in the library. But there are also other strands (or even excesses) of reference management, like the stories one could tell about the creator of the *Oxford English Dictionary*, James Murray; the philosopher Hans Blumenberg; Nietzsche's friend Franz Overbeck and his church encyclopedia; or the Viennese scholar Max von Portheim with his 500,000 entries on Josephinism—for the sake of brevity, they are omitted here.[2]

This much passion for ink and paper may come as a surprise in the digital age of *copy and paste* and hectically clicking keyboards, all the more if the index cards were compiled not only laboriously by hand, but *by one single person*. The high priests of reference management around 1900—for instance, Max von Portheim—obey the maxim of an influential reference theoretician of the eighteenth century: responding to a question about the actual authorship of the paper slips that form the basis for his more than one hundred monographs, Johann Jacob Moser mentions his fast-moving quill and discloses his trade secret: "Do it all yourself."[3]

Unlike Moser's solitary book production, this study was by no means written after the DIY maxim; on the contrary, without a lot of help and support it could barely have been written. Hence, for the German version of this study I owe a debt of gratitude to Peter Berz, Kenneth E. Carpenter, Susanne Holl, Friedrich Kittler, Thomas Macho, Harun Maye, Cornelius Reiber, Kei Saeki, Bianca Schemel, Margarete Vöhringer, Nikolaus Wegmann, and David Wellbery. Furthermore, the English edition would not have happened without the special intervention of Michael Buckland, hence my special thanks to him. Financial support for this translation was made possible by the Goethe Institute, Munich, as well as Samas GmbH, Worms. In addition, I owe Marguerite Avery, Wolfram Burckhardt, Anne Mark, Jasmin Meerhoff, W. Boyd Rayward, and Oliver Simons for their assistance and cooperation; last but not least, and particularly for the time in between the two books, thanks to Frau V.

Weimar, May 2010
Markus Krajewski

Notes

Chapter 1

1. From the catalog copy in figure 1.1.

2. Turing 1987, pp. 20, 91 and also in this connection, Dotzler 1996, esp. p. 7. For the structural analogy between card catalog and Turing machine, see above all note 15 in this chapter.

3. "For the duration of a state," as Müller 1995, p. 45 put it.

4. See Kittler 1986, pp. 244f., and Kittler 1993, pp. 170ff.

5. Translator's note: The author uses the German word *verzetteln* in a double sense: putting notes and quotes down on paper slips, and sorting individual notes in an index; but also disseminating, releasing them. See Grimm and Grimm 1956, cols. 2565ff. This ambiguity is dissolved in most cases by a translation decision that selects the former meaning and remains silent about the latter.

6. See Shannon 1948, pp. 6ff., and Kittler 1993, pp. 170f.

7. For this method (perhaps the original motivation for the present study), see Luhmann 1993.

8. When the famed library of Alexandria burns down in AD 640, the almost equally renowned *pinakes* of Callimachus are lost—a catalog comprising almost 120 volumes listing, describing, and sorting the Greek works and authors kept in the library. A *pinax* is mounted above the shelves to name authors and gather them in categories, establishing a basic systematic order. *Pinakes* offer a systematic order of knowledge. Although it is not known how Callimachus made his catalog, his work provides an order for long lists, by sorting alphabetically and by attempting systematic order in the service of a general classification of knowledge (in a library). This is presumably the first large-scale alphabetization of a library's holdings; in addition, each *pinax* represents a piece of a systematic order that even in its fragmentation represents a holistic

classification of knowledge in a library. Even if the *pinakes* were on papyrus, they could not be viewed as precursors of the card catalog because—owing to their materiality—the *pinakes* lack mobility. See Schmidt 1922 and Daly 1967 as well as, in general, Dawson 1962.

9. After the fourth century AD, Roman administration uses the *laterculum* (burned stone) as a register of all civil and military officers. As these usually consist of a continuous list of roles and a brief description of their responsibilities, and these roles are subject to change, the carrier medium itself must support this flexibility. Thus, the register consists of loose leaves held together by a clamp or a cord, allowing for additions and insertions; see Seeck 1924, pp. 905f. In contrast to Callimachus's catalog, which creates order only in copying mobile elements, with the *laterculi* what is noted can itself be moved. A second important characteristic is the necessity of creating an unambiguous link between carrier medium and idea or process—one man, one word; one office, one page. An idea or process must have its own carrier medium so it can be handled flexibly.

10. Foucault 1975, p. 363, n. 49; see Foucault 1966, pp. XV and passim for discourse analysis.

11. Reuleaux 1875, pp. 53f., emphasis in the original; for criticism of many other definitions of the machine concept, see pp. 592ff.

12. The last few decades have produced, starting with the work of Friedrich Kittler (1986), a specific direction in (German) media theory that applies Foucauldian discourse analysis and other methods of cultural study to develop a media history between epistemology and new forms of historiography; see the recent special issue of the journal *Grey Room: Architecture, Art, Media, Politics*, #29: "New German Media Theory" (MIT Press, Fall 2007).

13. Haraway 1997, p. 11.

14. Historians lament this, since the usually complex sources of historiography allow for many competing or complementary takes on the genealogy of a concept. For an exemplary treatment of computing myths, see Levy 1998.

15. This comparison is not to claim that the index catalog is already a Turing machine. Comparisons, transfers, and analogies are not that simple. If the elements of a universal discrete machine are present, they still lack the computational logic of an operating system, the development of which constitutes Turing's foundational achievement. What is described here is merely the fact that the card catalog is literally a paper machine, similar to a nontrivial Turing machine only in having similar components—no more, no less.

16. See for instance note 15 regarding the limits of the Turing machine analogy, note 1 of chapter 2 for the book flood metaphor, and notes 62 of chapter 2 and 42 of chapter 4 for the business card and bank note comparisons, respectively.

Chapter 2

1. See Giesecke 1998, pp. 158f., 171ff.; up to the Internet and its information overload, see Bickenbach and Maye 2009. See Bickenbach 1999, pp. 119–123, and Werle 2009, regarding the classical topos of the book flood or *multitudo librorum* since Jan Comenius's *Prodromus Pansophiae* (1637). In the seventeenth century, especially in the work of Johann Valentin Andreae and Jan Comenius, the mass of books is often cast in a nautical metaphor, in the context of navigation: a sea of scholarship, a flood of books. The metaphor carries a threatening and challenging semantics. Catalogs react to it by way of canalization (melioration), even if the flood of books is not fended off like water—it is not like a tsunami that recedes into the ocean. On the contrary, one usually wants to keep books. Nonetheless, books are not protected against misuse or neglect—be it to pave a road or to use them as kindling.

2. Doni 1550, cited after Zedelmaier 1992, p. 13.

3. Gardthausen 1920a, p. 3.

4. Buzás 1975, pp. 143f.

5. For biographical background, see Ley 1929, pp. 3–40.

6. Escher 1937, p. 119.

7. For the difference between bibliography and catalog from today's perspective, see Roloff 1961, pp. 244f., and Kittler 1979, p. 201.

8. On the question of why one might not count Johannes Trithemius with his *De Scriptoribus Ecclisiasticis* of the year 1494, see Zedelmaier 1992, p. 27, and Wellisch 1981, p. 10.

9. On the topical and Baroque rhetorical tradition, see Beetz 1980, pp. 125ff.

10. Gessner 1548, fol. 24, following Zedelmaier 1992, p. 88. It is worth noting the first hint about the origin of the German word *Kartei*: chart books, *chartaceos libros*, are apparently already in use around 1550.

11. The title pages of the *Bibliotheca Universalis* bear slightly varied depictions of a tree, mostly adorned by a reference (at times a giant frog) to the publisher, Christoph Froschauer.

12. The commentary "De Indicibus Librorum," for instance, begins with an elaborate preface, proving the importance Gessner accords to this point.

13. Zedelmaier 1992, pp. 63, 116.

14. Leyh 1929, p. 17.

15. Wellisch 1981, p. 13.

16. Wellisch (1981, p. 13) estimates the number of *locorum* per page at around 150, rightfully criticizing other authors' estimates of up to 1,000 entries per page.

17. Zedelmaier 1992, pp. 53f.

18. Gessner himself complains about his pale complexion and emaciated figure resulting from this activity; see Leyh 1929, pp. 21ff.

19. Gessner 1545, fol. 3; see Zedelmaier 1992, pp. 22f.

20. Gessner 1548, fols. 19f.

21. Gessner 1548, fol. 20; see also Zedelmaier 1992, p. 104.

22. Gessner 1548, fols. 19ff.; see also Wellisch 1981, p. 11.

23. Gessner 1548, fol. 20.

24. Gessner, 1548, fol. 19; see also Zedelmaier 1992, p. 105.

25. Gessner 1548, fol. 20.

26. Gessner 1548, fol. 20.

27. Zedelmaier 1992, pp. 103, 106. This suggests that books were cheaply available; otherwise, else Gessner would not have suggested cutting them up.

28. Schmidt-Künsemüller 1972, pp. 130, 134f.

29. Schmidt-Künsemüller 1972, p. 132.

30. Meyers Großes Konversations=Lexikon 1906, Art. "Buchdruckerkunst."

31. Gessner 1548, fol. 19, translated in Zedelmaier 1992, p. 103.

32. For his rich biography and further information, see Unterkircher 1968, pp. 81–99.

33. Hugo Blotius, preface to his first catalog, in Austrian National Library, Paper Manuscript Collection, HB Akt 2 1/2b/1576.

34. Unterkircher 1968, p. 109.

35. See Unterkircher 1968, p. 111 and above all p. 113. Work progresses slowly, because Blotius reads books very closely and fabricates ten or more paper slips on each: "Depending on the book design, some titles need the first or last name, some the content [...] The slips with titles were then cut up with scissors, so that each could be filed in different boxes depending on the alphabetical indexing of proper names or of topics."

36. Unterkircher 1968, p. 112.

37. On Jungius and his technique, see Meinel 1995.

38. Placcius 1689, p. 72.

39. For the later history of scrapbooks and notebooks and their contribution to the production of knowledge, see for instance te Heesen 2005; and for an extension of the copy-and-paste technique, see te Heesen 2006.

40. Placcius 1689, pp. 85–88.

41. Meinel 1995, p. 177.

42. Placcius 1689, pp. 121–158.

43. Placcius 1689, p. 155; see figure 2.7

44. Von Murr 1779, pp. 210f. Leibniz travels to Hamburg specifically to attend the auction of the library of Jungius's former student Martin Fogelio; see Lackmann 1966, p. 330.

45. For the cabinet in the history of science, particularly on its universal use in the taxonomic arrangement of minerals and the like, see te Heesen 2002.

46. Harsdörffer 1653, p. 57.

47. Around 1700, critics address the Baroque art of excerpting texts, with its *adversariis, florilegia*, and miscellanea that are usually not thematically sorted and accessible only via an index, as well as the systematic scrapbooks of collections, poetic cornucopias, and compilations of *locis communes* (compare Beetz 1980, pp. 144f., and Barner 1970, pp. 61, 232ff.). It is true that the polymaths give a positive twist to Baroque rhetoric and its poetological methods between 1690 and 1720 (Wiedemann 1967, p. 232), introducing a selective principle, *iudicium*, that cuts through the sheer accumulations of *memoria* and material collections (for details, see Beetz 1980, pp. 144–161). But the textual constitution of topical rhetorics—and of the materiality of scrapbooks—is superseded in the early Enlightenment by a procedure that banks on individual reading, emphasizing one's own access to sources and the use of one's own senses (Beetz 1980, pp. 146ff.). For a literary application, see Daniel Caspar Lohenstein's *Agrippina* (and with regard to this text, Kittler 1988, esp. pp. 50f.), as well as a mocking account of this transition in Lessing's *Der Junge Gelehrte Damis* (1748), and Wiedemann 1967 on that text.

48. See Löffler 1917, p. 100, and Lackmann 1966, p. 323, referring to Leibniz's requirement for being accepted among the ranks of the court counselors as a reason for his failure.

49. Milkau 1912, p. 593.

50. Löffler 1917, pp. 96ff.

51. See Scheel 1973, p. 188.

This catalog in fact was completed over the course of six years. Librarians Reinerding and Sieverds first copied the book titles onto folios on the basis of Duke August's catalogs between 1691 and 1695. Then library assistant Müller cut apart the thirty-two titles on each folio and ordered them alphabetically. Theology student Johann Clemens Müller then copied them into a book from 1695 to 1697, for two cents per folio. In 1697–98 he added a chronological index by publication year of the books. Finally, in 1698–99 the theology student Heinrich Balthasar Bergmann was hired to enter additions and supplements.

On Leibniz's suggestions about reordering the systematic shelving system, see elsewhere in this study.

52. Leibniz 1696, p. 71.

53. Jochum 1993, chap. 8.

54. Denis 1777, p. 277.

55. Leibniz 1699, p. 195. See also Löffler 1917.

56. Leibniz 1679, pp. 30f.

57. Leibniz 1699, p. 195.

58. See Leibniz 1699, p. 189. On project management in the age of Leibniz, see Krajewski 2004a.

59. Kayser 1790, p. 22.

60. Van Swieten 1780.

61. Bartsch 1780.

62. Anonymous 1795, p. 148. Reading academic and noble titles on small cards is analogous to accessing books one has not read, but knows nonetheless by their title and name. In contrast to handwritten catalog cards and bibliographies, which refer to all copies, visiting cards commonly exist in many uniform copies referring to one person. Individual index cards in a catalog, however, represent one identical library copy among a multitude of copies of that text. This inverse relation shifts a little once card printers begin circulating many identical cards for the many identical copies of a book, stimulated by Melvil Dewey in 1878 in relation to the intention of representing each book on an index card.

63. Goethe 1801, p. 454.

64. Jochum 2003, pp. 119f.

65. On the agents of this administrative act see again Jochum 2003, p. 119: for their supervisor Goethe, librarians check his own writings out from the collections and present them to him after having transferred their data carefully into catalogs according to his instructions.

Notes

Chapter 3

1. Hempel-Kürsinger 1826, p. 410.

2. Publication of the mayor's office and the Council of Vienna, January 7, 1771, Patente, H 45/1771, City Archive Vienna, cited in Wohlrab and Czeike 1972, p. 334. For a comprehensive history of conscription and house numbers in Europe, see Tantner 2007a,b.

3. Löffler 1956, p. 11.

4. Ebert 1820, pp. 14f., and see also Jochum 1991, pp. 15f., 20f.

5. Announcement of the mayor's office of the city of Vienna, January 7, 1771, H 45/1771, City Archive Vienna, cited in Wohlrab and Czeike 1972, p. 335.

6. Hempel-Kürsinger 1826, p. 410.

7. Hempel-Kürsinger 1825, p. 270.

8. Hempel-Kürsinger 1825, p. 270; see Aly and Roth 1984, on data management on punch cards after 1933.

9. Circular of the Austrian government "To the people of Vienna," September 9, 1792, cited after Wohlrab and Czeike 1972, p. 338.

10. Leyh 1961a, p. 729.

11. Leyh 1961a, p. 729; The equivalent terminology for the Staatsbibliothek in Berlin, Stiftung Preussischer Kulturbesitz: *Kriegsverlust*, casualty of World War II.

12. For example, Leyh (1961a, p. 731) does not conceal his anger when warning against idiosyncratic signature systems: "But where is the library user who would take care to learn this peculiar library language when a commonsensical catalog could direct him at once to the desired book with the aid of a simple number—that is, the *house number*" (emphasis mine).

13. See Jochum 1993, pp. 83 ff., and for the title as a first address Giesecke 1998, pp. 420ff.

14. Graesel 1902, p. 309.

15. Foucault (1966, p. 144) dates the catalog movement to the reorganization of libraries at the end of the classical period around the end of the eighteenth century.

16. Ebert 1820, pp. 16f.

17. Graesel 1902, p. 307.

18. Kayser 1790, p. 10.

19. Leyh 1961a, p. 693.

20. Rautenstrauch 1778, p. 172. The evident software command follows a deductive logic: the Latin numeral denotes a box, the Latin letter the drawer in the box, and the Arabic numeral the place of the book in the drawer.

21. Rautenstrauch 1778, p. 172.

22. Wegmann 2000, pp. 122f.

23. *Instruction, für die k. k. Universitäts- und Studienbibliotheken, provisorisch erlassen mit Stud. Hof-Comm.-Decrete vom 23. Juli 1825, Z. 2930*, in Grassauer 1883, p. 196.

24. Grassauer 1883, p. 197.

25. Meynen 1997, pp. 1f., 44.

26. "Among the abilities whose regular use is important to librarians, it may suffice to name memory here. As indispensable as a true and firm recollection of titles, names, and numbers is for the execution of a librarian's duties, it may be even more important to have a grasp of the local institutional memory; for a librarian who always has to consult a catalog to find what he needs is truly a pitiful man!" (Ebert 1820, pp. 14f.).

27. Kayser 1790, p. 3. Here Kayser turns out to be an important influence for Schrettinger.

28. For the complex history of the development of library signatures, see Leyh 1961a, pp. 688ff., 728ff.

29. The Danish traveler Werner Hans Fredrick Abrahamson describes the Wolfenbüttel library eight years after Lessing's death, in August 1788; see Wegmann 2000, p. 244.

30. See Wegmann 2000, pp. 245, 253ff., 258f. Lessing's successor as librarian promises a monetary reward to anyone who finds a word written by Lessing in the library; see Löffler 1917, p. 95.

31. Naudé 1627, pp. 97f.; for Leibniz, see Steierwald 1995, pp. 33ff.

32. See Schunke 1927, pp. 377ff., for a survey and genealogy of systems.

33. Leyh 1961b, p. 249.

34. Leyh 1921, p. 223.

35. See Rozier 1775, pp. v–viii, where the table produced by Rozier is discussed in detail, and pp. xi–xii on the playing cards that were used.

36. François 1974, p. 88. For exceptionally long titles one uses the ace; see figure 3.1 and Cole 1900, p. 330.

37. See the section "Revolution on Playing Cards" (which for chronological reasons comes later), describing the practical idea in the year 1791 to use playing cards in catalogs so as to compile information about all printed matter in France in a centralized national library in Paris.

38. For a detailed look at the debate between Martin Schrettinger's catalogs and Friedrich Adolf Ebert's systematics, see Jochum 1991 and Meynen 1997. Leyh (1914) is a polemical and decisive partisan of the end of systematic shelving. Yet according to Altmann (1928, p. 311), German libraries "almost without exception honor the idea of systematic shelving" even in 1928.

39. Austrian National Library, Vienna, Akt HB 125/1780, pt. 1.

40. Wieser 1968, pp. 240f.

41. On van Swieten's extraordinarily influential position at court, see Lesky 1973; for his medical training as a student of Herrmann Boerhaave in Leyden, see Lindeboom 1973.

42. Von Mosel 1835, p. 151. Apart from the continuous library metaphors about war and battle, the situation resembles that which confronted Hugo Blotius. One can only hazard a guess about this repeated coincidence of neglected libraries and Dutch scholars hired to clean them up. On the remarkable series of Dutch Catholics employed in the Vienna Court Library—from Hugo Blotius to his successor Sebastian Tengnagel to the van Swietens, father and son—see the Viennese exhibition catalog, Petschar 1993.

43. See the report by Gottfried van Swieten, son and successor to his father's office as prefect of the court library, recapitulating the previous two decades of the library's history: van Swieten 1787, p. 320.

44. Van Swieten 1787, p. 320.

45. See van Swieten 1787, p. 320; also see von Mosel 1835, pp. 192f., and Wieser 1968, p. 278.

46. Buzás 1975, p. 143. One is reminded of folios chained to desks, "loss through use," fires, and the lax morale of humanist scholars who took forever to return books, otherwise known as theft.

47. This extends from 1745 to 1803, interrupted only by the interim First Curator and later Director Kollar between Gerhard's death in 1772 and the new regime under Gottfried in 1777; see Stummvoll 1968.

48. For concrete numbers, see Bosse 1981, p. 85, and Schön 1987, pp. 41ff.

49. See Kittler 1995, pp. 37ff.,178ff.

50. Libraries keep having to regulate the appropriate delivery by printers since Blotius introduced the rule, as it is not always followed. The catalog of the book fair

in Leipzig is used as a permanent reference, since (for business reasons) it contains every new publication. See Wieser 1968, p. 278.

51. See Winter 1943, pp. 33ff., and Lesky 1973, pp. 19f.

52. See above all Lesky 1973, pp. 11–33.

53. Winter 1943, pp. 38ff. Also see Lindeboom 1973, p. 63; Lindeboom views Gerhard van Swieten's life as marked by Jesuit threats.

54. This is suggested by Bernhardt (1930, pp. 85f.), who develops a personality sketch of Gottfried van Swieten. See also Wangermann 1978.

55. Austrian National Library, Vienna, Akt HB 1780/134, cited after Radlecker 1950, p. 106.

56. Radlecker 1950, p. 107.

57. Von Mosel 1835, p. 154.

58. Von Mosel 1835, pp. 170f.

59. For this quotation and the following from Maria Theresa to Johann Thomas Trattner, see Bosse 1981, p. 10.

60. Von Mosel 1835, p. 183.

61. Wieser 1968, pp. 276f.

62. Gutkas 1989, p. 331.

63. Von Mosel 1835, p. 163.

64. Protocol of the court study commission, May 5, 1874, cited after Bernhardt 1930, p. 131.

65. For the education reform, see Bernhardt 1930, pp. 126ff.

66. Bernhardt 1930, p. 126.

67. See Winter 1943, p. 41, and in more detail Fournier 1876.

68. For a possible connection to Gessner's register, see Zedelmaier 1992, p. 100. Insight into censorship is offered in Petschar 1993, pp. 46–55.

69. Van Swieten, cited after Wieser 1968, p. 291.

70. Bernhardt 1930, p. 123.

71. Cahn 1991, *Der Druck des Wissens*, is an ambiguous title: it can be read as *The Pressure of Insight* or *The Printing of Knowledge*.

72. Letter from Gottfried van Swieten of October 24, 1781, Austrian National Library, Paper Manuscript Collection, Akt HB 1780/136.

73. Von Mosel 1835, p. 151.

74. Van Swieten 1787, p. 320. See also Petschar, Strouhal, and Zobernig 1999, as well as Meinel 1995, p. 183, n. 57, or Roloff 1961, pp. 255, 257.

75. Cataloging according to oral tradition is practiced in large libraries all the way into the twentieth century (e.g., in Tübingen and Darmstadt); see Hilsenbeck 1912, pp. 313ff. For the earlier practices, see also Schreiber 1927.

76. Austrian National Library, Vienna, Akt HB 125/1780.

77. Austrian National Library, Vienna, Akt HB 125/1780, pt. 1.

78. The instructions follow a definite *if-then* structure; see Bartsch 1780, pp. 125f.:

I. Must [...] Should [...] then would [...]
II. If one finds [...] then one must [...]
III. If there are [...] then [...]
IV. If one [...] then there is [...] ...

79. Bartsch 1780, p. 125.

80. Petschar 1999, pp. 24f.

81. Von Mosel 1835, p. 177.

82. Today's coding standard (see Hagen 1993, pp. 151f.) does not allow the working algorithm to jump—as in older programming languages like BASIC with its famous GOTO command. In this sense, the process of creating a complete catalog on the basis of index cards is hardly *basic*.

83. Denis 1777, pp. 274f.

84. Petschar 1999, p. 28. Starting in 1835, the library no longer employs reader's assistants, as von Mosel reports with regret (1835, p. 177, n. 2). See also Förstmann 1886 and Krajewski 2010.

85. Bartsch 1780, p. 131.

86. Their form again simulates the design of a book (see figure 3.2). Thus, a card catalog borrows the appearance of a bound catalog. This camouflage is supposed to calm down traditional librarians, but I have not tracked down its inventor. For its later use in Gießen, see Haupt 1888.

87. Austrian National Library, Vienna, HB Akt 126/1780, pt. 2.

88. Bartsch 1780, p. 132.

89. Kayser (1790, p. 49, emphasis mine) therefore admonishes:

> The more one works through repositories, the more cards accumulate, of course. As soon as they become too numerous, it is advisable to sort them alphabetically into heaps and to arrange them on a table. If the table is in a place where no draft or any other accident can disarrange

the paper slips , then I consider it a suitable storage for the slips, more so than a cabinet. For I have to consult the stacks far too often. On the table they are all readily at hand. If they are in a cabinet, I have to take them out and search for them anew, which causes *time loss*."

90. Milkau 1912, p. 604.

91. Van Swieten 1787, p. 320.

92. Grassauer 1883, p. 169.

93. Jesinger 1926, p. 12. For a statistical overview of the gains, see Dosoudil and Cornaro 1994, p. 182. The influx ebbs around 1815.

94. For his biography, see von Wurzbach 1873, pp. 67–69, von Schulte 1888, and Werner 1888, p. 155.

95. See Rautenstrauch, *Was bei der akademischen Bibliothek bereits gemacht worden*, May 13, 1777, cited after Jesinger 1926, p. 19.

96. Rautenstrauch 1778, p. 173, point 7. And anticipating this, Denis 1777, p. 260.

97. 1. Systematic division of books by an order of knowledge, with theology coming first and philology last. 2. Shelving of books by format, not too closely together. 3. Addressing by shelf, row, and box. See Rautenstrauch 1778, pp. 171f.

98. Rautenstrauch 1778, p. 172.

99. See Hittmair 1901, pp. 9ff., on the instructions of 1778 and their successor set of the year 1825, especially in comparison to practices in other countries.

100. The basic catalog is extant in the holdings of the university library; see Hainz-Sator 1988, p. 117.

101. For the assumption in 1836 that this completes the project, see Ladewig 1912, p. 211. Firm belief in the indestructible paradigm of closure and absolute order is evident also when, in 1886, the Royal Library in Berlin considers the paper slips as a mere interim solution; see Ermann 1908, p. 464.

102. Winter 1943, p. 194.

103. Cited after Jesinger 1926, p. 37.

104. Jesinger 1926, p. 83, n. 16.

105. See Dosoudil and Cornaro 1994, pp. 201f. Regarding a lasting index on paper slips, I also spoke to Leopold Cornaro, in Vienna, myself.

106. Hopkins 1992, p. 379.

107. See Cole 1900, p. 329, with remarks on decrees for protecting books from dust and dirt, and similar instructions about relics, gold and silver, paintings, and sculptures.

Notes

108. Hopkins 1992, p. 383.

109. Cole 1900, p. 330.

110. Letter from Gaspard-Michel LeBlond to La Rochefoucauld, December 5, 1790, cited after Riberette 1970, p. 17.

111. Hopkins 1992, p. 387.

112. Hopkins 1992, p. 391.

Chapter 4

1. Placcius 1689 marks a peak in the literature on the techniques of excerpting texts. Yet the discourse is extended—for instance, by Schmeizel (1728, pp. 176ff.), who in the tradition of the history of scholarship offers huge apparatuses of annotation with synoptic illustrations. See also Bertram 1764.

2. This function is mirrored in the etymology of the word "catalog," as enumeration or stock taking. See Löffler 1956, p. 11.

3. Flusser 1996, p. 32.

4. Bogeng 1915, pp. 138f., emphasis mine. The comparison is hardly an accident, given Gessner's plant books—and address books in cities, in Vienna perhaps? *Quod erat demonstrandum.*

5. The early practitioner of indexing, Joachim Jungius, already perfects the idiosyncrasy of his machine. Three days after his death, one of his students begins a careful autopsy of his index cards. There are no aids for access, no apparatus; neither signatures nor a numbering of the cards, neither registers nor indexes, let alone referential systems that guide one to the building blocks of knowledge. See Meinel 1995, pp. 172f.

6. Translator's note: the word *diskret* in German homonymously denotes both "discreet" in the sense of tactful, courteous, and "discrete" in the sense of separate, distinct.

7. Ebert 1820, pp. 21f.

8. See Kittler 1979.

9. Enzensberger 1995, p. 31: "Etwas Zerfetztes im Minenfeld, / daneben ein unverwundeter Schuh, / Flöße in der Karibik / alles kommt über Satellit, / wird gespeichert d.h. vergessen." ("Something shredded in the mine-field / next to it, a shoe, intact / floats in the Caribbean / everything comes via satellite / stored, i.e. forgotten").

10. See Luhmann 1993, and see Krajewski 2011 (in press) for an attempted critical analysis.

11. Vismann 2000, pp. 88f.

12. The paper slips of the academic machine, however, are fixed on square pieces of wood and this can be turned only on two axes, limiting the slips' freedom; see Swift 1728, pp. 102f.

13. Moser 1773, pp. 42f.

14. Moser 1773, pp. 50ff.

15. Moser 1773, p. 54.

16. Moser 1773, p. 41. For a bibliography of all of Moser's texts, except for the theological works, see the appendix to a text by the librarian Albrecht Christoph Kayser from 1790.

17. The cultural constant afforded by the box of paper slips as the basis for text production maintains its currency. "Hagen: This architecture of slips, this dimension of boxes is huge by now, isn't it?—Luhmann: It is pretty extensive, yes. Hagen: Several meters. Luhmann: Yes, yes. Hagen: ... and that is the basis, as it were, for your work. Luhmann: Yes. Hagen: Without it ... if this was taken away, it would be hard. Luhmann: Yes, that would make things difficult" (Hagen 2004, p. 107).

18. As a brief illustration, not far removed topically as it appears under the title "A Small Footnote to the Philosophy of the Writing Surface," one might refer here to what Moser notes during his five years of internment in Hohentwiel Castle, with a view of Lake Constance:

> I was allowed neither paper, nor ink, nor a quill, nor a pen, and I had no books but the Bible and the Steinhofer sermons, later a songbook as well [...] I would have loved to compose spiritual songs, but how? [...] I found that I could scratch into the white wall with the tip of my lamp-cleaner, and much was gained! In the beginning it was crude and large; later I learned to make it smaller and finer. Now I wrote all over the wall in my room, as far as I could reach. Yet now it was on the wall, but how would I take it with me when I was discharged? (Moser 1777, p. 144 ff.)

One might append here a small media history of the increasing elasticity and mobility of the writing surface, with protagonists from Moser's stony and immobile surface, to Gerhart Hauptmann's wallpaper (he "wrote his ideas at night onto the wallpaper near his bed"; Kunert 1986, p. 394), all the way to Arno Schmidt's small but mobile card catalog (Schmidt 1995, p. 31).

19. Moser 1777, pp. 102–108.

20. Compare the commentary by Eduard Berend (in Jean Paul 1796a, p. 294), or Stockhammer 2000, pp. 53f., esp. n. 39. Indeed, one finds in Jean Paul's excerpts a reference to Moser's *Autobiography (Lebensgeschichte)*, yet bibliographically incorrect: "Moser, Johann Jacob: Lebensgeschichte, von ihm selbst beschrieben. 4 Tle., Offenbach 1768" (Müller 1988, p. 143). The 1768 first edition has neither a place of publication nor four parts, and in particular, it has no §46 nor any description of

Notes

Moser's way of working. Thus, the excerpt must actually refer to the 1777 second edition.

21. Jean Paul 1796a, pp. 15, 91.

22. Jean Paul 1796b, p. 771.

23. For the competition between index card and book, see more below in this study.

24. Jean Paul 1796b, p. 771.

25. For a description of excerpt notebooks and their development, see Müller 1988 and Böck 2001.

26. Meinel 1995, p. 170; see also Birus 1986, p. 50, and compare Böck 2002.

27. Moser 1773, pp. 58f.

28. Jean Paul 1796b, p. 772.

29. Birus 1986, p. 50; see also Böck 2002.

30. The attribute "learned" or "scholarly" includes Jean Paul and his excerpts, even though his contemporaries explicitly challenged any association of learning or scholarship with him and his texts, which he took as a real insult; see Birus 1986, p. 51.

31. Jean Paul 1796b, p. 771. Elsewhere, he writes about card games, "Thus, the question is whether it would not be pleasing and useful to have a collection of essays that mixes ideas from all sciences not like so many poisons but like playing cards, so as to benefit those who know how to win at such a game, like the one with Lessing's dice; of course, when it comes to such a collection, I have one and increase it every day, be it only to render my mind as free as the heart should be" (Jean Paul 1804, pp. 202f., n. 1).

32. Moser 1773, p. 64.

33. For another, more elaborate discussion of completeness, see Krajewski 2006.

34. Rosenkranz 1844, pp. 12f.

35. Moser's autobiography is published in three editions between 1768 and 1777, and in 1789 is supplemented with an index, securing quite a broad reception. In Hegel's youth, Moser was one of the most widely read legal scholars, so we can assume that this reading was also entered into Hegel's collection of excerpts.

36. Regardless of whether Hegel's famous collection of index cards might still lie undiscovered in some Berlin attic, as an employee of the Stiftung Preußischer Kulturbesitz suspects—thus posing an unsolved problem as to where to address research questions—or whether it is indeed lost, one can still gain an impression of Hegel's practice of excerpting. Besides various handwritten notecards that already indicate the desired path of the young student—*academy, pedagogy, scholarship,*

knowledge of the Egyptians—Hegel's manuscripts from his time in Stuttgart also contain entries about monks, the soul, and the "path to happiness in the larger world"; see Hoffmeister 1936, p. 100.

37. Kittler 1997, p. 197.

38. Heine 1826, pp. 284f.

39. Curtius 1954, p. 57; see Goethe 1831, p. 305: Mephisto citing the Bible, Eph. 6:12.

40. Deutsche Bundesbank 1953, p. 1800. On later Prussian promissory notes, see Pick 1967, p. 45.

41. Born 1972, p. 4.

42. Paper money and index cards only fulfill their purpose if they exhibit a specific mobility, when they can circulate and be reshuffled. Yet while bank notes are interchangeable (two 100 Reichsmark notes always represent the same value in gold), index cards in a catalog are not. For two different index cards represent, not the equivalent value, but individual data—separate books or different information. Both modes of representation are nonetheless closely related as soon as identical catalog cards are produced centrally, when printers toward the end of the nineteenth century produce large sets of the same card. Conversely, there are situations where paper money can be perceived as unique and referring to highly singular elements: for instance, in the rare case of police registration of a certain serial number to withdraw it from circulation or solve a crime. These interlaced logics of representation of knowledge management on paper slips and economic management with banknotes are related independently of these limited cases, as becomes evident in frequent cross-references from library management to finance by Leibniz or Goethe. For example, Goethe (1801, p. 454) wrote that in the library he "felt the presence of great capital that silently yields incalculable interest" (more on this later in the present study).

43. Paradigmatic here: Gosch 1789.

44. Foucault 1966, pp. 196f.; see also Beer 1845, p. 11. One might note an irony of history in that the library at the Viennese court and its mobile catalogs, established for two decades, are struck just as severely: starting in 1808 they are funded by increasingly inflationary banknotes, which in 1809 lead to a budget collapse and to significant difficulties in managing the books; see von Mosel, p. 220. Also see Born 1972, pp. 5f.: as a reaction to these disturbances, the famous Peel's Act was passed in 1844, monopolizing the issue of paper money and elevating the banknote to legal tender, thus to money in the strict sense.

45. Born 1972, p. 16. It was only in 1848 that Prussian bank law permitted the issue of paper slips and banknotes.

Notes 161

46. For warnings, see Block 1807 for example. Though there had always been regulations before banks were allowed to issue paper, they appear to have been too rudimentary to prevent abuse. For measures taken to enhance the effectiveness of the limitations, see Perrot 1874, p. 6.

47. Bartsch 1780, pp. 131f.

48. Schopenhauer 1819, p. 71.

49. Stadermann 1994, p. 38.

50. Beer 1845, p. 16.

51. Beer 1845, p. 16.

52. "Where my borrowings are concerned, see whether I have been able to select something which improves my theme: I get others to say what I cannot put so well myself" (Montaigne 1588, p. 458).

53. See for instance the historical perspective of a catalog that lists citations, from the American pioneers of citation analysis: Lehnus 1974.

54. See the commentary under the pleasant but slightly misleading title "The Interest of Data Bankers: Does the Scholar's Republic Need an Antitrust Agency for Citations?" (Küster 2000).

55. Beer 1845, p. 7.

56. Beer 1845, p. 7. For the juxtaposition of circulating blood and circulating ideas, see Schmidt and Sandl 2002.

57. Müller 1977, p. 94.

58. For this setting of the cult of genius, see Schneider 1994, pp. 82f.

59. Grafton 1990.

60. For this "pride in collections" is appropriate only for "gray mice born by the mountain"; see Meyer 1907, p. 13.

61. Examples of literary attempts that obviously use the index card as status symbol and authorial signature are Schaukal 1913, 1918, as well as Borchert 1985 and Hildesheimer 1986 at the dawn of the personal computer.

62. Michel 2002, pp. 38ff. For the history of the *Encyclopedia* as a knowledge system, see Darnton 2001.

63. Luhmann 1993, p. 58.

64. Luhmann 1993, pp. 59f.

65. Krajewski 2011, in press.

66. No doubt there is a long list of possible supporters for this thesis, from Konrad Gessner in the mid sixteenth century, to Bertolt Brecht or Martin Heidegger, to contemporary sociologists of science like Bruno Latour.

67. Kleist 1805, p. 405.

68. Kleist 1805, p. 406.

69. See Kant 1797, 2-II. Doctrine of the Methods of Ethics, 1. § 50, on the relation between student and teacher: "the midwife of his thoughts."

70. Luhmann 1993, p. 57.

71. Kleist 1805, p. 408.

72. See Schmidt 1995, and the macroscopically detailed illustrations in Ruetz 1993. See also the motif of index cards in Nabokov's *Pale Fire*, not to mention the fifty paper slips of Nabokov's last, unfinished novel, *The Original of Laura*; see Krajewski 2001.

Chapter 5

1. Letter from Harvard president John Thornton Kirkland to John Davis, July 28, 1812. Unless otherwise specified, materials cited in the notes for this chapter come from the Harvard University Archives.

2. Letter from William Croswell to the Harvard Corporation, November 10, 1824.

3. On Croswell's entrepreneurial career—for instance, as editor of *A Mercator Map of the Starry Heavens* or as private tutor of an American captain's children in Liverpool and London—see Lovett 1963, pp. 5–18.

4. Letter from William Croswell to the Harvard Corporation, April 15, 1829.

5. Letter from William Croswell to the Harvard Corporation, August 1833.

6. Turner 1978, p. 88.

7. Stored in the Harvard University Archives, HUG 1306.5.

8. In December 1812, the diary documents real progress: "Fr. 17. Begin Italian / Th. 24. Finish 10. Alcove / Fr. 25. Begin 30 Alcove."

9. Letter from William Croswell to Judge Davies, March 8, 1821.

10. Letter from William Croswell to the Harvard Corporation, November 10, 1824. For more details about Jacques Charles Brunet's classification scheme, especially in comparison to other contemporary concepts, see Schunke 1927, p. 386.

11. Letter from William Croswell to President Kirkland, March 10, 1817.

12. The majority of Croswell's complaints are articulated in letters only after his dismissal. On June 6, 1821, for instance, he writes to the Harvard Corporation, "There was injury from the Fruit Boys during the first years. Their station was on the Southern side of Holden. I requested in vain they might stand on the Northern side. I frequently left the Library on account of their piercing shrieks. More might be added. It is owing to the use of the cold Bath, that I am still, in some measure, capable of business." In a letter of October 25, 1819, Croswell offers a compressed version of the disturbances: "I must now complain. In July and August I was disturbed by Fishermen, blowing their horns violently; by persons firing guns, and otherwise. It was in vain to keep my windows shut in the hottest weather. The consequence was numbness in my head, with dizziness, and restless nights for two months."

13. Letter from William Croswell to the Harvard Corporation, October 25, 1819.

14. Draft of a letter by William Croswell, dated August 6, 1833.

15. Letter from William Croswell to Judge Davies, February 1821. In his next letter to the judge, in the following month, Croswell shows signs of paranoia. He suspects nothing but foul play: "An unconnected Plan was given to me adapted to perplex and discourage and wear me out. [...] It was clothed with such authority that I followed it scrupulously. Some of my repetitions may be charged to it" (letter from William Croswell to Judge Davies, March 8, 1821).

16. Letter from William Croswell to the Harvard Corporation, November 10, 1824.

17. Letter from William Croswell to the Harvard Corporation, November 10, 1824.

18. Indeed, the printed catalog of 1790 paves the way for library economies: Croswell has no scruples about cutting printed pages into small scraps and gluing them together to make a "new" catalog (figure 5.2). "The Catalogue of 1790 has been a valuable guide," he writes elsewhere (letter of October 25, 1819), evidently a grateful commentary on the idea of saving time and effort with the aid of paper slips.

19. Letter from William Croswell to President Kirkland, December 13, 1820.

20. Today, Harvard owns two copies of Gessner's *Pandectae*. However, as both editions were acquired only in the first half of the twentieth century, Croswell had no opportunity to read Gessner's recommendations.

21. Letter from William Croswell to Judge Davies, January 18, 1821. Of these volumes, twenty-one are extant and kept in the Harvard University Archives.

22. Letter from William Croswell to President Kirkland, March 8, 1821.

23. Letter from William Croswell to President Kirkland, September 17, 1827.

24. See the letter from William Croswell to Judge Davies, January 18, 1821.

25. Letter from William Croswell to President Kirkland, September 17, 1827.

26. The announcement also refers to Croswell's many years of experience: "It is proper to add, that I have spent some years in Bibliographical pursuits."

27. For reasons of conservation, the solubility of the glue could not be tested at the Harvard University Archives. Konrad Gessner had recommended an easily soluble glue to allow paper slips to be moved around in a hybrid book; see figure 2.5.

28. See Ticknor 1874, pp. 57f. On Goethe's library activities, see Jochum 2003.

29. See Turner 1978, p. 88. For the bustling activity of the American community in Göttingen, see Long 1935.

30. Leyh 1921.

31. Walton 1939, p. 29.

32. Walton 1939, p. 29.

33. Turner 1978, p. 89.

34. Walton 1939, p. 30.

35. Thaddeus William Harris, "Ninth Annual Report," 1840, Harvard University Archives.

36. Walton 1939, p. 34.

37. Ebert 1820, pp. 14f.

38. Abbot 1862, pp. 3f.

39. Abbot 1862, p. 11.

40. Abbot 1864, p. 39.

41. Abbot 1864, pp. 39f.

42. Abbot 1862, p. 9.

43. Abbot 1864, p. 38.

44. Cutter 1869, p. 99.

45. See Harsdörffer and Schwenter 1636, pp. 523f., and in this study, passim.

46. Cutter 1869, p. 99.

47. For details, see Abbot 1862, pp. 8ff., and 1864, pp. 107ff.

48. Abbot 1862, p. 12.

49. Sibley 1863, p. 32.

50. Potter and Bolton 1897, p. 41.

51. On Sibley's achievement in this context, see Mitchell 2003.

52. See Folsom 1864. Compare Abbot's (1867, p. 1) summary of the last interim state: "At present [April 1867], about half of the books in the Library have been entered upon it."

53. Cutter 1869, p. 128.

Chapter 6

1. Vann 1978, p. 9.

2. Utley 1926, p. 12. On Dewey's understanding of "best reading," see Wiegand 1996, pp. 88f., 130.

3. The many exceptions students are forced to learn claim unnecessary attention and complicate the language, which in turn makes it harder for foreigners and children to learn. Thus, Dewey later hires teachers to calculate that between two and four years of education could be saved with a rationalized spelling reform and exclusive use of the metric system; see Wiegand 1996, pp. 33, 43.

4. For Dewey, as a defender of the metric system, it has to be December 10, 1879, his own birthday, when he metamorphoses "Dewey" into "Dui"; see Wiegand 1996, p. 63.

5. Dawe 1932, p. 252.

6. Diary entry, March 7, 1873, cited after Wiegand 1996, p. 20.

7. Dawe 1932, p. 182.

8. Melvil Dewey, *Library Classification System*, with the remark "Original idea of May 8, 1873, Submitted to Library Com. of Amherst College," cited after Dawe 1932, pp. 31f. See also the fully elaborated system in Dewey 1911. Also see Lackmann 1966, emphasizing the proximity of the Dewey decimal classification system to Leibniz's development of a universal library.

9. This desired global reach is followed around 1900 in Europe by vehement criticism not only against classification (see Niemann 1927, p. 10), but also against the claim for its distribution throughout America's libraries (see Milkau 1898, p. 20).

10. Bührer and Saager 1912, p. 6. For the ambitious attempt to disseminate Dewey's ideas in Germany, see "The Bridge Enters the Office: World Brain" in chapter 7.

11. Bührer and Saager 1912, p. 8.

12. Wiegand 1996, p. 33.

13. One possible cause for Dewey's time-saving craze is offered by his biography. In winter 1868, fire breaks out in the Amherst College library. Dewey tries to rescue as many books as possible, contracting a severe case of smoke poisoning. A doctor gives

him just two more years to live, thus establishing in Dewey's mind the idea of an excessive lack of time; see Wiegand 1996, pp. 10f.

14. This strange habit of bookkeeping is already manifest in Dewey's childhood diaries; see Wiegand 1996, pp. 8, 14.

15. Wiegand 1996, pp. 39, 53.

16. Melvil Dewey in a letter to Justin Winsor, June 27, 1876, cited after Wiegand 1996, p. 42.

17. Wiegand 1996, p. 42.

18. Wiegand 1996, p. 42; for the "time savings" see also Wiegand 1996, pp. 48f.

19. Scott 1976, p. 295.

20. American Library Association 1877, p. 285.

21. Datz 1926, p. 669.

22. Brown 1894, p. 49.

23. American Library Association 1877, pp. 285f. The standard format of today's American postcards differs a bit: it is 8.9 × 12.7 cm.

24. Siegert 1993, p. 148.

25. For the format debates in Brussels, see Junker 1896 and Field 1896.

26. Hilsenbeck 1912, p. 312.

27. Wiegand 1996, p. 59.

28. Wiegand 1996, p. 64.

29. Wiegand 1996, pp. 66f.

30. With little success, as the new business goes bankrupt in July 1882; see Wiegand 1996, p. 74, n. 27.

31. Melvil Dewey, letter dated March 20, 1881, cited after Wiegand 1996, p. 71.

32. As a vector anchored in the library and pointing to the coordinates of the "office."

33. Cutter and Bowker 1888, p. 96.

34. Melvil Dewey in a letter to his lawyer, S. J. Elder, January 21, 1888, cited after Wiegand 1996, p. 111.

35. Wiegand 1996, p. 113.

36. Sherman 1916, p. 42.

37. See Flanzraich 1993, pp. 406, 421, as well as Flanzraich 1990, p. 295.

38. Library Bureau 1909, pp. 3ff.

39. Brown 1894, p. 50.

40. James 1902, p. 187.

41. See Flanzraich 1993, p. 406, and Library Bureau 1909, p. 42.

42. This repeatedly unsuccessful project is pursued again later, thanks to Dewey's influence, yet always remains a loss leader; see Scott 1976, pp. 297, 300, and Wiegand 1996, p. 238. In Germany, the debate over centralized printing of index cards finally begins at the end of the 1890s, aiming at a general catalog for the German Reich. See Milkau 1898 and Berghoeffer, Bess, and Schultze 1906.

43. Herkimer County Historical Society, Library Bureau papers, *Record of Salesmen's Meetings, 1898–1900*, cited after Flanzraich 1993, p. 406.

44. *The Quarter Century Club of Library Bureau*, cited after Flanzraich 1993, p. 407.

45. Wiegand 1996, pp. 234f.

46. It is only in 1898 that the opposing directors manage to disrupt Dewey's influence in favor of public libraries by transferring control of the enterprise into eight hands rather than just Dewey's own two. This restructuring and its profit-oriented guidance finally banish Dewey's influence for good. After selling his shares successively after 1902, he leaves the company; see Wiegand 1996, pp. 239ff.

47. Wiegand 1996, p. 236; see also Dewey 1888.

48. The flow of capital during the second half of his life will allow Dewey to found the Lake Placid Club in Albany, New York, and later a similar venture in Florida. On these projects and their relation to Dewey's hay fever, see Vann 1978, p. 44, and Dawe 1932, p. 231.

49. Wiegand 1996, pp. 111, 235.

50. "The great feature which has caused librarians the world over to count the card catalog as *the greatest library invention*, is the ease of keeping it up to date and in perfect order" (Davidson and Parker 1894, p. 27, emphasis mine).

51. Davidson and Parker 1894, p. 3.

52. Davidson and Parker 1894, p. 28.

53. Davidson and Parker 1894, p. 27, emphasis mine.

54. Davidson and Parker 1894, p. 28.

55. Davidson and Parker 1894, p. 3.

56. Thus, the marketing strategy increasingly relies, apart from independent fabrication and production, on individual delivery of standardized products. Before office furniture is installed in a company, an analysis of the client's work process—a kind of early management consulting—prepares for the unavoidable transition to the new optimizing and rationalizing techniques of the efficiency movement. This modern service is offered under the title "Improved Business Methods":

> It [Library Bureau] offers its services to any business house, agreeing after examination to attain a certain saving in expenses and to maintain or increase efficiency. The proved value of such services is the basis for the fee charged. It studies the unproductive side of the expense account and considers after carefully studying the details of each business how it is possible to reduce expenses by changes in method or adoption of new devices. The Bureau brings to the questions involved not only a large and varied experience among offices and business houses, but the services of the most expert accountants and students of business methods. (Davidson and Parker 1894, p. 7)

57. Davidson and Parker 1894, p. 34.

58. Library Bureau 1902, p. 95.

59. Herkimer County Historical Society, Library Bureau papers, *Record of Salesmen's Meetings, 1898–1900*, cited after Flanzraich 1993, p. 411.

60. For the historical genesis of upright storage of files, see Vismann 2000, pp. 79f., 129ff., 142f.

61. See Library Bureau 1903 and Schellenberg 1961, pp. 80ff., as well as Flanzraich 1993, pp. 416f. On the hanging file system in general, see Yates 1989, pp. 56–63.

62. Bertillon 1895.

63. Flanzraich 1993, p. 414.

64. See Flanzraich 1990, pp. 96ff., and Flanzraich 1993, p. 415.

65. Herbert Davidson, "Confidential Statement Concerning Library Bureau," September 1, 1907, Columbia University Archives, New York, Melvil Dewey Papers (MDP), Box 100.

66. Letter from Melvil Dewey to Herbert Davidson, February 7, 1908, Columbia University Archives, MDP Box 64. Also see Dewey 1887.

67. See Brown 1894, p. 49 and passim.

68. Dewey has Library Bureau honor him as the inventor of the card catalog; see Sherman 1916, p. 42 ("inventor of the card system").

69. Letter from Melvil Dewey to Hosea Paul, March 6, 1916, Columbia University Archives, MDP Box 64. Remarkably, Dewey also claims to have invented the "loose leaf ledger idea."

70. Letter from Burt C. Wilder to Melvil Dewey, December 12, 1912, Columbia University Archives, MDP Box 39.

71. Letter from Melvil Dewey to Burt C. Wilder, December 16, 1912, Columbia University Archives, MDP Box 39. Dewey seems to know of Abbot's acts at Harvard College, since Library Bureau published an essay by Harvard president Charles William Eliot in the in-house journal *Bureau Drawer*, vol. 1, no. 6, December 1907, p. 2.

72. Jewett 1853, p. 24.

73. See Conner 1931, pp. 51f., Utley 1951, and Anonymous 1915.

74. The first patent of this sort, however, was issued in 1887 to Joseph Fezandie; see Flanzraich 1993, pp. 409f. Other concepts can be traced back to 1636, to Harsdörffer and Schwenter 1636, pp. 523f.; see Klinckowström 1934. Despite his European roots, the secretary of the Ohio YMCA, James Newton Gunn, claims the invention of *tabs*, although they can already be found in the work of Johann Jacob Moser; see also Frank 1922. Gunn secures his patent 1896, sells it to Library Bureau, and serves the company as a consultant; see Flanzraich 1993, p. 412. His management consulting brings him to the attention of the greatest patent patron of the time, the General Electric Company; see Davidson and Parker 1894, p. 66. In 1902, the journal *World's Work* portrays Gunn flatteringly as a new type of "production engineer"; see Smith 1902.

75. See Flanzraich 1993, p. 409. Pierced cards can be found in Ezra Abbot's catalog construction of 1861. Abbot may have gotten the idea from Charles Folsom, who presented his security suggestions at the 1853 librarians' conference in New York:

> [It] consisted of a series of cards, about nine inches long and two wide, which were laid in a pile, and a hole bored through each end of the whole, and strings passed through them. These strings were of such a length as to allow the whole of the cards to be slid back or forward, as the writer or compositor should find necessary, yet still preserving them in their proper order, without confusion or danger of loss. The whole were fitted into a box of the requisite size, from which they could be drawn singly, without deranging the consecutiveness of each. (Utley 1951, p. 83)

Doubts about cards must have been widespread around 1850; only Folsom's experiments at the Boston Athenaeum instill trust in the materiality of index cards, and even these cards maintain the odd format that Thaddeus William Harris took from William Croswell's paper slips. "The cards are long and narrow and are kept in cases, made to resemble folio volumes, one side of which opens like the cover of a book" (Jewett 1853, p. 31). This demonstrates the competition between book and card index, so that the latter will go under cover, literally, to hide its alternative structure.

76. Baker 1994, pp. 64f.

77. See Ernst 2003, and against this legend, Cortada 1993, p. 45.

78. Remington Arms Company 1999, Unisys Corporation 1999, and Flanzraich 1993, p. 405.

79. Schneider 2007.

80. Petschar, Strouhal, and Zobernig 1999, pp. 14, 164f.

81. Wiegand 1996, p. 371.

Chapter 7

1. Wiegand 1996, p. 54.

2. Ladewig 1912, p. 209.

3. For the repeated debate over whether the bound book is the better catalog form after all, see Keysser 1885, p. 2, and later Holst 1937. With reference to xerographic copying, see Roloff 1967.

4. This will be discussed in the chapter 8.

5. In his famous essay "The Question Concerning Technology" from 1954, Martin Heidegger uses the word "Ge-stell" as an abstract concept (which is outlined below); literally this word means "library shelf" as well as "frame."

6. Leyh 1929, p. 2.

7. "Whereas the Halle stacks from 1878 or the Strasbourg stacks from 1889 are like a heavily armed hoplite, the 1910 Tübingen stacks resemble an elegant fencer" (Leyh 1929, p. 14).

8. Graesel 1902, p. 138, and Leyh 1929, p. 11.

9. Heidegger 1954, p. 21.

10. Heidegger 1954, p. 20.

11. Kaiser 1921, p. 107. See Hortzschansky 1908 for the history of the catalogs in Berlin.

12. Kaiser 1921, p. 103.

13. Leyh 1929, p. 11.

14. Ladewig 1912, p. 252.

15. Ladewig 1912, pp. 254f.

16. "The uniform beauty of the index card cabinet can be perceived even if it starts small," as it says on p. 257 of *Politics of the Library*. Combinatory boxes and recursive mobility: even the boxes themselves, not only the drawers and cards in them, are mobile units and multiply in their recombinations.

17. Ladewig 1912, p. 255.

18. Von Morzé 1982, p. 392.

19. Müller 1986, p. 245.

20. Roloff 1961, p. 247.

21. Von Leon 1820, p. 49.

22. Musil 1932, p. 502.

23. Ladewig 1912, p. 218.

24. Roloff 1961, p. 247.

25. Ladewig 1912, p. 255.

26. The "slit system" is an alternative to a hole in the lower part of the card pierced by a thin round rod. Instead there is a notch (not a hole) in the shape of a "T" in each card. The rod can be twisted so that cards can be added or removed without needing to remove the rod.

27. An example is found in the Bonn University library—the asymmetrical stick that turns to free or lock the cards, fabricated by the Friedrich Soennecken company; see Ermann 1926.

28. Ladewig 1912, p. 255.

29. Compare the relatively early security system by Aristide Staderini, who provided cards with a hole at the bottom that was held by an additional wooden staff; see Staderini 1896, as well as Graesel 1902, pp. 257–271. Even earlier ones are described in Molbech 1833, p. 144; see the synopsis in Gardthausen 1920b, pp. 40f.

30. Ladewig 1912, p. 256.

31. Kaiser 1921, p. 108.

32. See Kaiser 1921, p. 103; for ruminations on further measures against theft, see also Ermann 1926.

33. See Flanzraich 1990, pp. 89–93. In 1896, a Paris branch is added; see pp. 96ff. Later, there are branches in Cardiff, Birmingham, and Manchester, but none in Germany.

34. Regarding this dispute, see Hofmann 1916, esp. pp. 10f., 46f., criticizing Ladewig's adoption of the American public library movement with its distinction of good and bad literature. See the same work regarding Hofmann's concept of the popular library. Ladewig 1917 counters with the public library.

35. Von Morzé 1982, p. 393. For a more detailed history of this dispute, see Stieg 1986.

36. La Fontaine 1903. For the memorial function of the project, see Ernst 2003; and for Otlet's pioneering role in documentation and informatics, see Buckland 1997.

37. Thron 1904, p. 11.

38. Thron 1904, p. 5.

39. Junker 1897, p. 7.

40. See Junker 1897, p. 7, and Thron 1904, p. 20.

41. Thron 1904, p. 12.

42. Junker 1897, p. 7.

43. Niemann 1927, p. 5.

44. In the internal bulletin of the Brussels institute, these images are accompanied by artistic organigrams in an aesthetics that is taken up by other adaptations of worldwide registration, for instance, in Dewey's decimal classifications.

45. Bührer and Saager 1911, p. 2.

46. See Bührer and Saager 1911, p. 1. That a bridge takes you from one side of a stream to the other side eludes the metaphor here.

47. Bührer and Saager 1912, p. 11. For a more comprehensive description of the project, see Krajewski 2006, chap. 2.4, pp. 102–129, as well as Hapke 1997 and Sachsse 2004.

48. Saager 1921, p. 8. Around 1910, the addressee, Wilhelm Ostwald (1853–1932), is not only a Nobel Prize–winning founder of the field of physical chemistry but also a theoretician of organization.

49. "As he said, his plan was by and large done. But as his expression appeared to others to be unclear and jumpy, it was impossible to understand his plans without devoted attention, and so I developed a small program and then wrote a book, once a publisher was found, in which I attempted to sift through and present Bührer's thoughts, systematically ordered and theoretically founded." Amid the Munich bohemians he meets a remarkable person, "a prophetic nature with strangely formulated ideas of salvation that certainly had a sound core, even if one did not, like their prophet, see them as the center of the world—a pugnacious fanatic for his idea for which he was willing to make any sacrifice, whom one might forgive his occasionally unwonted ardor—in short, a fascinating character who had the right to more attention than most of the countless apostles who haunted Munich. [Adolf Hitler perhaps?] While the other prophets all preached of health, soul, culture, humanity, religion, and other ideals, this man wanted to salvage the planet by the most prosaic and banal means—the paper format" (Saager 1921, p. 2).

Notes

50. See the illustration after the cover page of Ostwald 1933. Polemically satirizing this was Carl Schmitt, writing under the pseudonym Johannes Negelinus and Mox Doctor (1913, p. 17).

51. Die Brücke 1911b, p. 1.

52. See Ostwald 1912b, pp. 17f. In addition to this project, which no by accident reminds one of Dewey's reform, The Bridge distributes his idea of decimal classification; see Bührer and Saager 1912, pp. 3f.

53. See Ostwald 1933, p. 295, and Bührer and Saager 1911, pp. 88f. The latter text is also the starting point for the publication series by The Bridge and the first application of the *Monographic Principle*.

54. Ostwald 1933, p. 295.

55. Ostwald 1933, p. 295.

56. This mode of knowledge generation exhibits surprising isotopic parallels to the notational practice of systems theory; see Luhmann 1993, and the commentary in Krajewski 2011, in press.

57. See Ostwald 1933, p. 301, and Krajewski 2006.

58. See the series edited by Ostwald, *Classics of the Exact Sciences*, and later *Great Men: Studies on the Biology of Genius*; see Pinner 1918.

59. Saager 1911, p. 9.

60. The cards are kept in world format without room for marginalia, for they are the least conducive "for noting mental participation in the content. Rather, all notes, ideas, comments, references, etc., that arise in studying belong on index cards that are also expected to be in world format, so that they can be entered in the card index for later use" (Ostwald 1912c, p. 2). Secondary world formats thus allow recursive storage of what has been stored. The original series of cards (A series or normal format) requires a box (B series) that is slightly larger and contained in a drawer (C series) that in turn is slightly bigger... Only the postcard is still left out and has yet to be integrated.

61. Die Brücke 1911a, p. 4.

62. The English brochures have titles like "The Newest Force in Business Building" or "Armorclad Guides" and carry pictures of warships (both Library Bureau, no year).

63. Hottinger 1911.

64. Hottinger 1911, p. 2.

65. Hottinger 1911, p. 3.

66. Hottinger 1911, p. 3.

67. Bührer and Saager 1911, p. 18. See also Krajewski 2006.

68. Meyer 1907, p. 4.

69. Meyer 1907, p. 14.

70. Meyer 1907, p. 9.

71. For example, an advertisement of the Hinz Fabrik in Berlin-Mariendorf carries the slogan "The Hinz customer database—the memory of your business!" in *Das System* 1 (5), p. 129. For the memorial discourse of storage systems, see Ernst 2003.

72. Meyer 1907, p. 10.

73. Meyer 1907, p. 11.

74. Meyer 1907, p. 14.

75. Bührer and Saager 1911, pp. 26f.

76. Ostwald 1933, pp. 291, 293.

77. Saager 1921, p. 36. In the beginning, the letters from Bührer to Ostwald show the usual devotion; see the correspondence in the Ostwald estate in the archive of the Berlin-Brandenburg Academy of Sciences.

78. See May 1997, Gal 1991, pp. 9f., and Krajewski 2004b.

79. Saager 1921, p. 12. Note the importance of water in all these projects.

80. Bührer and Saager 1911, p. 18.

81. Bührer and Saager 1911, p. 20.

82. Ostwald 1933, pp. 291, 293.

83. Saager 1921, p. 46.

84. Saager 1921, p. 11.

85. See Gödel 1931 and, for an exegesis of his theorem of incompleteness, Hofstadter 1989, pp. 93f., 110ff.

86. Meyer 1907, p. 13.

87. Ostwald 1933, p. 304.

88. Ostwald 1933, p. 305.

Chapter 8

1. For unplanned use of the card index in Berlin, see Ermann 1908, pp. 464f.; and for its description as a sanctum preserved only to initiated librarians, see Ippel 1916,

Notes 175

pp. 29f. For the librarians' attempts of systematizing World War I in advance, see Buddecke 1913, and for the history of its instantaneous cataloging, see Berz 1993.

2. For the energetic imperative, see Ostwald 1912a, pp. 1ff., 81ff.

3. Burchardt 1977, p. 72, and Rabinbach 1992, p. 254.

4. Roesler, introduction to Taylor 1913, pp. 5f.

5. Burchardt 1977, p. 60, displays an amorous caricature from *Life* magazine 1913 with the caption, "Young man, are you aware that you employed 15 unnecessary motions in delivering that kiss?"

6. Popp 1932, pp. 13f.

7. Witte 1925, p. 29. The same author secures her place as distributor of the American office reform movement in Germany, for instance with Witte 1930.

8. Haußmann 1925, p. 4.

9. Roth-Seefrid 1918, p. 18.

10. Ostwald 1929, p. 232.

11. Popp 1932, p. 10.

12. Schlesinger 1920, pp. 21f. Motion study has a longer tradition and was certainly not Taylor's invention. Jean-Baptiste Colbert, Prussian generals, and Charles Babbage had already conducted studies to optimize motion sequences; see Burchardt 1977, p. 57.

13. For their pioneering role, see Yates 1989.

14. Giebichenstein 1929, p. 26. For the conceptual history of "rationalization," see Hinnenthal 1927; and in relation to the watchword "efficiency," see also Berg 1929.

15. Frank 1922, p. 7.

16. For its American forebears and its German implementation, see Witte 1930, pp. 55–58, and Witte 1926.

17. Burchardt 1977, p. 69.

18. For details, see Vismann 2000, pp. 132ff.; and for the description of a customer registry, see Bjd 1900.

19. Flanzraich 1993, p. 412.

20. Scheffler 1928, p. 35.

21. Frank 1922, p. 3.

22. Frank 1922, p. 3.

23. Anonymous 1914, pp. 2, 122–127.

24. Soennecken 1914, p. 119.

25. Osterwald 1914, p. 253.

26. The BUGRA, which is underway when the war breaks out, quickly expands its scope and, and it turns itself into the "First German War Exhibition" by adding weaponry and field hospital cabinets to the graphic arts and office administration tools. See Schramm 1914, p. 277.

27. Haußmann 1925, p. 5.

28. *System* was merged in 1929 with *Das Geschäft*, and in 1930 the two became *Beleg und Bilanz*. The oldest of these German journals was founded in 1898 under the simple name *Organisation*. No doubt *Wirtschaftlichkeit* was named in allusion to the imperial office *Reichskuratorium für Wirtschaftlichkeit*; see Hinnenthal 1927, pp. 19f.

29. Haußmann 1925, p. 46.

30. Regarding the translation of the concept "organization" as "regulation," see also the neologistic attempts by Porstmann (1928a, p. 295), associating index cards with train logistics, as if to please the fictional General Stumm.

31. Dux 1922, p. 13.

32. Herrmann 1931, p. 8.

33. See the hymn to paper slips, with a detailed listing of their advantages, in Le Coutre 1931.

34. Breiter 1927.

35. Menzel 1926, p. 30.

36. Le Coutre 1931.

37. For this attitude to book keeping in a literal sense, see Schellenberg 1961, p. 248. See also his counterarguments concerning lack of space, lack of ability to provide an overview, and lack of context, requiring cross-references.

38. See Vogel 1931, p. 61: "The soul of the view card index is in the messages it sends."

39. Vogt 1922, p. 25.

40. See Rabofsky 1932.

41. Dux 1922, p. 6.

42. Dux 1922, p. 45.

43. Weise 1928.

Notes

44. Le Coutre 1931.

45. For the notion of "modernity," see Gumbrecht 1978.

46. Müller 1986, p. 249. See Kittler 1998 on the "power of the desk."

47. Anonymous 1931.

48. Giebichenstein 1929, p. 27.

49. Le Coutre 1931, p. 55.

50. Warneyer and Koppe 1927, p. 57.

51. Hummel 1928, p. 161.

52. See Flanzraich 1993, p. 413. For the renewed "all-American" reinvention of loose-leaf accounting, see Stoeckel 1946, and for its further dissemination, Wootton and Wolk 2000.

53. Herrmann 1931, p. 8; on this topos, see also Weise 1928.

54. See Weise 1928, p. 79, and of course Osterwald's automatic accounting, "which no longer allows for mistakes" (Osterwald 1914, p. 253).

55. For details, see Mühlpfordt 1908, Hummel 1928, and Weise 1928.

56. For a literary studies contribution to the question of how a book could otherwise look, see the debate around 1800 between Friedrich Schlegel and Friedrich von Hardenberg (see Schreiber 1983) or the suggestions by Stéphane Mallarmé (see Ingold 1993).

57. Mühlpfordt 1908, p. 100.

58. Commentary following Mühlpfordt 1908, p. 102; see figure 8.3.

59. Fickert 1926, p. 4, as well as Paulukat 1910; also see Vismann 2000, p. 19, for a discussion of the metaphorical implications of blockage in files.

60. Anonymous 1929d. The German word *Nominalstil* refers to a style in which there is a preponderance of nominal constructions.

61. Bohne 1930, p. 73.

62. Herrmann 1931, p. 11.

63. Porstmann 1928b, p. 360.

64. Anonymous 1929a, p. 247.

65. Anonymous 1929c; see also Vogt 1922.

66. As Porstmann called himself in the third edition of *Karteikunde*. See Porstmann 1939, p. 12.

67. Porstmann 1928b.

68. See chapter 2, especially note 10, of this book for more detail.

69. For a brief history of mercantile recording techniques since antiquity, see Porstmann 1928, p. 9: "This is the origin of the card index. There have always been paper slip arrangements; the medieval trader worked with them as well as the Greeks and Romans. The Roman trader carried a number of wax tablets on his belt to write on. These individual pages are a precursor of the book and of the card index." End of story.

70. Anonymous 1848, p. 141.

71. "The German word *Kartei* ['card index'] has not gained traction, unfortunately. If our office supply industry used the word in its literature, its career would soon be guaranteed" (Frank 1922, p. 3).

72. Porstmann 1928b, p. 361. Solidarity is also declared in Anonymous 1929b.

73. Anonymous 1929c.

74. Porstmann 1928b, p. 360.

75. Porstmann 1928b, p. 359. Compare the suggested replacement of *Lose-Blatt-Buch* ("loose-leaf book") by *Blattei* and of *Hängeregistratur* ("hanging folder file") by *Mappei* in Porstmann 1929. According to Drosdowski 1993, the *-ei* suffix goes back to a linguist named Pfister in 1889. Later, the universal collective noun *Datei* ("data file") will innocuously join this paradigm.

76. Porstmann 1939, p. 12.

77. Breiter 1951, p. 118.

78. Porstmann 1928a, p. 13.

79. Wegmann 2000, pp. 47f.

80. Porstmann 1928a, p. 12. The connotations of the term indeed enjoy popularity in esoteric applications; see Esdorp 1934.

81. Jochum 1993, pp. 158f.

82. For organization as its own (meta)science, see Ostwald 1929, p. 232.

83. Anonymous 1929e.

84. Porstmann 1928a, pp. 255f. For lack of space, I will not repeat the long history of this transition; see Cortada 1993, esp. pp. 44–63, 128ff.

85. Bohne 1930, p. 75.

86. Ostwald 1914, p. 119.

Notes

87. Vogt 1922, p. 82, emphasis in the original. Porstmann, who served as Wilhelm Ostwald's assistant from 1912 to 1914 and was thus familiar with his ideas from the times of The Bridge, is named as the editor of this text, and calls his 1928 sequel *(Karteikunde)* a second edition. The fact that the passage quoted here does not appear in the second version indicates that it was Vogt's.

88. Vogt 1922, p. 83.

89. Vogt 1922, p. 85.

90. Porstmann 1928a, p. 327; see also Meynen 1998, p. 82. On hypertext, see Vilém Flusser's analysis of the dissolution of books as a result of digital technologies (Flusser 1987), as well as Deleuze and Guattari 1977, p. 10.

91. The German verb *anzetteln* first denotes the preparatory framing of a loom; see Pfeifer 1995, p. 1605. In the eighteenth century, it mostly functions as the equivalent of *instigating*, not least fights and wars; see Grimm and Grimm 1854, vol. 1, col. 526.

92. Benjamin 1928, p. 456.

93. Vogt 1922, p. 83.

94. Vogt 1922, p. 84.

95. Predeek 1927, p. 469.

96. Benjamin 1928, p. 456.

97. Fickert 1926, p. 9.

98. Yet the metaphor of the *Kartei* or card index is present more than ever in graphic user interfaces; see Meynen 1998, p. 86.

99. For early "successes," see Aly and Roth 1984.

100. Hohn 1993, Kieser 1997, and Neuland 1995.

101. Strouhal 1999, p. 15.

102. Gardthausen 1920b, p. 141.

Afterword

1. Luhmann 1987, p. 143.

2. Winchester 2003; Buchberger et al. 2007.

3. Johann Jacob Moser 1777, §46, pp. 102f.

References

Abbot, Ezra. 1862. To the Honorable, the President and Fellows of Harvard College. Unpublished letter, Harvard University Archives, Cambridge, Mass.

Abbot, Ezra. 1864. Statement respecting the New Catalogues of the College Library. In *Report of the Committee of the Overseers of Harvard College Appointed to visit the library for the year 1863*, ed. Harvard College Library, 35–68. Boston: Press of Geo. C. Rand & Avery. Submitted July 10, 1863.

Abbot, Ezra. 1867. *Brief Description of the Catalogues of the Library of Harvard College*. Cambridge, Mass.: Printed at the Library.

Altmann, Wilhelm. 1928. Zur Aufstellung und Katalogisierung der Bücher. In *Von Büchern und Bibliotheken*, ed. Gustav Abb, 310–315. Berlin: Verlag von Struppe & Winckler.

Aly, Götz, and Karl Heinz Roth. 1984. *Die restlose Erfassung: Volkszählen, Identifizieren, Aussondern im Nationalsozialismus*. Berlin: Rotbuch Verlag.

American Library Association. 1877. Preliminary Report. *American Library Journal* 1:283–286.

Anonymous. 1795. Neueste Moden in Visiten-Karten. *Journal des Luxus und der Moden* 10:147–150.

Anonymous. 1848. Der Hofbibliotheksakt zum Katalog von 1848: Instruction für die mit der Leitung der außerordentlichen Arbeiten an der k. k. Hofbibliothek beauftragten Beamten und Hülfsarbeiter. In *Der Zettelkatalog: Ein historisches System geistiger Ordnung*, ed. Hans Petschar, Ernst Strouhal, and Heimo Zobernig, 139–144. Vienna: Springer Verlag.

Anonymous. 1914. "Der Kaufmann" auf der buchgewerblichen Weltausstellung Leipzig 1914, *Die Bugra. Leipzig 1914. Mitteilungen von der Internationalen Ausstellung für Buchgewerbe und Graphik*, no. 7, 2.

Anonymous. 1915. *Proceedings of the Librarian's Convention held in New York City, September 15, 16 and 17, 1853*, Reprinted for William H. Murray, Cedar Rapids, Iowa.

Anonymous. 1929a. Das Wort "Kartei"—Gut der Allgemeinheit. Nicht mehr Privileg einer Einzelfirma. *Zeitschrift für Organisation* 3 (9):247.

Anonymous. 1929b. Der Streit um das Wort Kartei. *Rationelle Betriebsführung*, n.p.

Anonymous. 1929c. "Kartei" als Warenzeichen gelöscht! *Rationelle Betriebsführung*, n.p.

Anonymous. 1929d. Loseblatt-Buchführung. *Wirtschaftlichkeit* 53, n.p.

Anonymous. 1929e. Zur Anwendbarkeit der Lochungskarten in der Buchführung. *Büro-Organisation* 3 (4):31.

Anonymous. 1931. Frißt die Büromaschine Seelen? *Büro-Organisation* 5 (10):73.

Baker, Nicholson. 1994. Discards. *New Yorker* (4):64–86.

Baker, Nicholson. 2001. *Double Fold: Libraries and the Assault on Paper*. New York: Random House.

Barner, Wilfried. 1970. *Barockrhetorik: Untersuchungen zu ihren geschichtlichen Grundlagen*. Tübingen: Max Niemeyer Verlag.

Bartsch, Adam. [1780] 1999. Einige Bemerkungen die Verfertigung eines neuen Catalogs der gedruckten Bücher in der k. k. Bibliothek betreffend. In *Der Zettelkatalog: Ein historisches System geistiger Ordnung*, ed. Hans Petschar, Ernst Strouhal, and Heimo Zobernig, 125–138. Vienna: Springer Verlag.

Beer, Wilhelm. 1845. *Bemerkungen über Zettel-Banken und Papiergeld*. Berlin: Gebrüder Unger.

Beetz, Manfred. 1980. *Rhetorische Logik: Prämissen der deutschen Lyrik im Übergang vom 17. zum 18. Jahrhundert*. Vol. 62 of *Studien zur deutschen Literatur*. Tübingen: Max Niemeyer Verlag.

Benjamin, Walter. 1928/1996. One-Way Street. In *Selected Writings. Vol. 1, 1913–1926*, ed. Marcus Bullock and Michael W. Jennings, 444–488. Cambridge, Mass.: Belknap Press.

Berg, Karl. 1929. Efficiency. *Rationelle Betriebsführung*, n.p.

Berghoeffer, Ch. W., B. Bess, and W. Schultze. 1906. Gesamtkatalog und einheitlicher Zetteldruck. *Zentralblatt für Bibliothekswesen* 23 (2):53–66.

Bernhardt, Reinhold. 1930. Aus der Umwelt der Wiener Klassiker. Freiherr Gottfried van Swieten (1734–1803), *Der Bär. Jahrbuch von Breitkopf & Härtel auf die Jahre 1929/1930*:74–166.

Bertillon, Alphons. 1895. *Das anthropometrische Signalement: Lehrbuch der Identifikation von Verbrechern, Angeklagten oder Verhafteten, von Verunglückten, Selbstmördern etc.*

References

2nd enhanced edition. Bern: A. Siebert. Authorized German edition, ed. Dr. Ernst von Sury.

Bertram, Philipp Ernst. 1764. *Entwurf einer Geschichte der Gelahrtheit für diejenigen, welche sich den schönen Wissenschaften, der Weltweisheit und der Rechtsgelehrsamkeit widmen*. Halle: Bey Johann Jusinus Gebauer.

Berz, Peter. 1993. Weltkrieg/System: Die 'Kriegssammlung 1914' der Staatsbibliothek Berlin und ihre Katalogik. *Krieg und Literatur* 5 (10):105–130.

Bickenbach, Matthias. 1999. *Von den Möglichkeiten einer 'inneren' Geschichte des Lesens*. Tübingen: Max Niemeyer Verlag.

Bickenbach, Matthias, and Harun Maye. 1999. Endlich surfen! Eine Poetik nasser Medien. *Sprache im technischen Zeitalter* 37 (50):221–234.

Birus, Hendrik. 1986. *Vergleichung: Goethes Einführung in die Schreibweise Jean Pauls*. Vol. 59 of *Germanistische Abhandlungen*. Stuttgart: J. B. Metzlersche Verlagsbuchhandlung.

Bjd. 1900. Kunden-Register. *Organisation: Fachblatt der leitenden Männer in Handel und Industrie* 2 (9):89.

Block, Georg Wilhelm. 1807. Ueber die Errichtung einer Zettel- und Leihebank für die Hannöverischen Lande. *Minerva: Ein Journal historischen und politischen Inhalts* 62 (May):282–303.

Böck, Dorothea. 2002. Die Taschenbibliothek oder Jean Pauls Verfahren das "Bücher-All" zu destillieren. In *Masse und Medium: Verschiebungen in der Ordnung des Wissens und der Ort der Literatur 1800/2000*, ed. Inge Münz-Koenen and Wolfgang Schäffner, 18–40. Berlin: Akademie Verlag.

Bogeng, Gustav Adolf Erich. 1915. *Streifzüge eines Bücherfreunds. Zweiter Teil*. Vol. 21 of *Ordentliche Publikationen der Gesellschaft der Bibliophilen*. Weimar: Gesellschaft der Bibliophilen.

Bohne, Ernst Rudolf. 1930. Die neue Sachlichkeit in der Buchführung. *Büro-Organisation* 4 (10):73–75.

Borchert, Jürgen. 1985. *Mein mecklenburgischer Zettelkasten: Aufenthalte und Wanderungen*. Rostock: VEB Hinstorff Verlag.

Born, Karl Erich. 1972. *Die Entwicklung der Banknote vom Zettel zum gesetzlichen Zahlungsmittel*. Wiesbaden: Steiner Verlag.

Bosse, Heinrich. 1981. *Autorschaft ist Werkherrschaft: Über die Entstehung des Urheberrechts aus dem Geist der Goethezeit*. Vol. 1147 of *UTB*. Munich: Ferdinand Schöningh.

Breiter, Josef Robert. 1927. Welche Kartei-Typen gibt es? *Wirtschaftlichkeit* 2 (14):n.p.

Breiter, Josef Robert. 1951. Dr. Walter Porstmann 65 Jahre! *Zeitschrift für Organisation* 20 (4):118.

Brown, James Duff. 1894. Mechanical Methods of Displaying Catalogues and Indexes. *Library* 6 (62):45–60.

Brücke, Die. Internationales Institut zur Organisierung der geistigen Arbeit. 1911a. *Satzung*. 3rd edition. Munich: Dr. C. Wolf & Sohn Kgl. Hof- und Universitäts-Buchdruckerei.

Brücke, Die. 1911b. *Was "Die Brücke" will*. Munich: Selbstverlag der Brücke.

Buchberger, Reinhard, Gerhard Renner, and Isabella Wasner-Peter, eds. 2007. *Portheim—sammeln & verzetteln: Die Bibliothek und der Zettelkatalog des Sammlers Max von Portheim in der Wienbibliothek*. Vienna: Sonderzahl.

Buckland, Michael K. 1997. What Is a "Document"? *Journal of the American Society for Information Science* 48 (9):804–809.

Buckland, Michael K. 2006. *Emanuel Goldberg and His Knowledge Machine: Information, Invention, and Political Forces*. Vol. 37 of *New Directions in Information Management*. Westport, Conn.: Libraries Unlimited.

Buddecke, A. 1913. Zur Systematik des Kriegswesens. *Centralblatt für Bibliothekswesen* 30 (12):538–544.

Bührer, Karl Wilhelm, and Adolf Saager. 1911. *Die Organisierung der geistigen Arbeit durch "Die Brücke."* Ansbach: Fr. Seybold's Buchhandlung.

Bührer, Karl Wilhelm, and Adolf Saager. 1912. *Die Welt-Registratur: Das Melvil-Deweysche Dezimal-System*. Ansbach: Fr. Seybold's Buchhandlung.

Burchard, Gustav. 1880. Professor Burchards internationaler uniformer Catalogzettel. *Neuer Anzeiger für Bibliographie und Bibliothekswissenschaft* 686:221–224.

Burchardt, Lothar. 1977. Technischer Fortschritt und Wandel: Das Beispiel der Taylorismus-Rezeption. In *Deutsche Technikgeschichte: Vorträge vom 31. Historikertag am 24. September 1976 in Mannheim*, ed. Wilhelm Treue, 52–77. Vol. 9 of *Studien zur Naturwissenschaft, Technik und Wirtschaft im Neunzehnten Jahrhundert*. Göttingen: Vandenhoeck & Ruprecht.

Buzás, Ladislaus. 1975. *Deutsche Bibliotheksgeschichte des Mittelalters*. Vol. 1 of *Elemente des Buch- und Bibliothekswesens*. Wiesbaden: Dr. Ludwig Reichert Verlag.

Cahn, Michael. 1991. *Der Druck des Wissens: Geschichte und Medium der wissenschaftlichen Publikation*. Wiesbaden: Reichert Verlag.

Cole, George Watson. 1900. An Early French "General Catalog." *Library Journal* 25:329–331.

References

Conner, Martha. 1931. *Outline of the History of the Development of the American Public Library, Preliminary Edition*. Chicago: American Library Association.

Cortada, James W. 1993. *Before the Computer: IBM, NCR, Burroughs, and Remington Rand and the Industry They Created, 1865–1956*. Princeton, N.J.: Princeton University Press.

Curtius, Ernst Robert. 1954. Goethes Aktenführung. In *Kritische Essays zur europäischen Literatur*, 57–69. Bern: Francke.

Cutter, Charles Ami. 1869. The New Catalogue of Harvard College Library. *North American Review* 108 (222):96–129.

Cutter, Charles Ami, and Richard R. Bowker. 1888. The Library Bureau Failure. *Library Journal* 13 (3/4):96.

Daly, Lloyd W. 1967. *Contributions to a History of Alphabetization in Antiquity and the Middle Ages*. Vol. 90 of *Collection Latomus*. Brussels: Latomus, Revue d'Études Latines.

Darnton, Robert. 2001. Eine kleine Geschichte der Encyclopédie und des enzyklopädischen Geistes. In *Die Welt der Encyclopédie*, ed. Anette Selg and Rainer Wieland, 455–464. Frankfurt am Main: Die Andere Bibliothek, Eichborn Verlag.

Datz, Harry R. 1926. A Pioneer: The Library Bureau. *Library Journal* 51:669–670.

Davidson, Herbert E., and W. E. Parker. 1891. *Classified Illustrated Catalog of the Library Bureau: A Handbook of Library and Office Fittings and Supplies*. Boston: Library Bureau.

Davidson, Herbert E., and W. E. Parker. 1894. *Classified Illustrated Catalog of the Library Bureau: A Handbook of Library and Office Fittings and Supplies*. Boston: Library Bureau.

Dawe, Grosvenor, ed. 1932. *Melvil Dewey, Seer: Inspirer: Doer. 1851–1931*. Lake Placid, N.Y.: Lake Placid Club, Essex Co.

Dawson, John M. 1962. A Brief History of the Technical Services in Libraries. *Library Resources & Technical Services* 6 (3):197–204.

Deleuze, Gilles, and Félix Guattari. 1977. *Rhizom*. Berlin: Merve Verlag.

Denis, Michael. 1777. *Einführung in die Bücherkunde*. Vienna: Bey Johann Thomas Edlen von Trattnern.

Deutsche Bundesbank. 1953. *Deutsches Papiergeld 1772–1870*. Munich: Typographisches Institut Giesecke & Devrient.

Dewey, Melvil. 1887. Evolution of the Card System. *Library Notes* 2:29–34.

Dewey, Melvil. 1888. The Library Bureau. *Library Journal* 13 (5):145–146.

Dewey, Melvil. 1911. *Decimal Classification and Relativ Index for Libraries, Clippings, Notes, etc.*, 7th. Ed. Essex, N.Y.: Lake Placid Club.

Doni, Anton Francesco. 1550/1551. *Libraria.* Vol. 2. Venice.

Dosoudil, Ilse, and Leopold Cornaro. 1994. Universitätsbibliothek Wien: Hauptbibliothek. In *Handbuch der historischen Buchbestände in Österreich.* Vol. 1, Vienna, part 1, ed. Österreichische Nationalbibliothek, 177–207. Hildesheim: Olms-Weidmann.

Dotzler, Bernhard J. 1996. *Papiermaschinen: Versuch über Communication & Control in Literatur und Technik.* Berlin: Akademie Verlag.

Drosdowski, Günther, ed. 1993. *Duden "Das große Wörterbuch der deutschen Sprache."* Vol. A–Bim. 2nd completely revised and expanded edition. Mannheim: Dudenverlag.

Dux, Wilhelm. 1922. *Die Kartei des Kaufmanns.* Vol. 27 of *Biolets Globus-Bücherei.* Stuttgart: Verlag von Wilhelm Biolet.

Ebert, Friedrich Adolf. 1820. *Die Bildung des Bibliothekars.* 2nd revised edition. Leipzig: Steinacker & Wagner.

Enzensberger, Hans Magnus. 1995. *Kiosk: Neue Gedichte.* Frankfurt am Main: Suhrkamp Verlag.

Erman, Wilhelm. 1908. Die Königliche Bibliothek zu Berlin: Ihre Geschichte und ihre Organisation. *Centralblatt für Bibliothekswesen* 25 (10):463–465.

Erman, Wilhelm. 1926. Zur Anordnung und Aufbewahrung alphabetischer Zettelkataloge. *Centralblatt für Bibliothekswesen* 43:554–556.

Ernst, Wolfgang. 2003. *Im Namen von Geschichte: Sammeln, Speichern, Er/Zählen. Infrastrukturelle Konfigurationen des deutschen Gedächtnisses.* Munich: Wilhelm Fink Verlag.

Escher, Hermann. 1937. Konrad Gessner über Aufstellung und Katalogisierung von Bibliotheken. In *Mélanges offerts à M. Marcel Godet, Directeur de la Bibliothèque Nationale Suisse à Berne à l'occasion de son soixantième anniversaire,* 119–127. Neuchâtel: Paul Attinger.

Esdorp, Viktor. 1934. *Kartothek des Ich: System einer Lebensführung.* Vienna: Saturn-Verlag.

Fickert, M. 1926. *Taylorisierte Behörden-Buchhaltung mit 4 Muster-Formularen: Ein gehobenes kameralistisches System in Konteiform.* Stuttgart: Taylorix Organisation.

Field, Herbert Haviland. 1896. Das geeignetste Format der bibliographischen Zettel. *Bulletin de l'Institut international de bibliographie* 1:202–204.

Flanzraich, Gerri Lynn. 1990. The Role of the Library Bureau and Gaylord Brothers in the Development of Library Technology, 1876–1930. Doctoral dissertation, Columbia University.

Flanzraich, Gerri Lynn. 1993. The Library Bureau and Office Technology. *Libraries & Culture* 28 (4):402–429.

References

Flusser, Vilém. 1987/2002. The Future of Writing. In *Writings*. Trans. Erik Eisel, 63–69. Minneapolis: University of Minnesota Press.

Flusser, Vilém. 1996. *Kommunikologie*. Vol. 4 of *Schriften*. Mannheim: Bollmann Verlag.

Folsom, Charles. 1864. Report. In *Report of the Committee of the Overseers of Harvard College Appointed to Visit the Library for the Year 1862: Together with the Accompanying Documents*, ed. Harvard College Library, 3–14. Boston: Press of Geo C. Rand & Avery.

Förstmann, Ernst. 1886. Die Bibliotheksdiener. *Centralblatt für Bibliothekswesen* 3:190–196.

Foucault, Michel. 1966/2002. *The Order of Things: An Archaeology of the Human Sciences*. London: Routledge.

Foucault, Michel. 1969/1995. *Archäologie des Wissens*. 7th edition. Frankfurt am Main: Suhrkamp Verlag.

Foucault, Michel. 1975/1998. *Überwachen und Strafen: Die Geburt des Gefängnisses*. 12th edition. Frankfurt am Main: Suhrkamp Verlag.

Fournier, August. 1876. Gerhard van Swieten als Censor: Nach archivalischen Quellen. *Sitzungsberichte der Akademie der Wissenschaften Wien* 84 (3):387–466.

François, André. 1974. *Histoire de la carte à jouer*. Serge Freal, Ivry.

Frank, Max. 1922. *Die Kartei*. Berlin: W. F. Marten.

Gal, E. 1991. Geschichten vom Finden. *Schattenlinien* 2 (4/5):3–35.

Gardthausen, Victor. 1920a. *Handbuch der wissenschaftlichen Bibliothekskunde. Erster Band*. Leipzig: Verlag von Quelle & Meyer.

Gardthausen, Victor. 1920b. *Handbuch der wissenschaftlichen Bibliothekskunde. Zweiter Band*. Leipzig: Verlag von Quelle & Meyer.

Gessner, Konrad. 1545. *Bibliotheca Universalis, sive Catalogus Omnium Scriptorum Locupletissimus, in Tribus Linguis, Latina, Graeca, & Hebraica*. Zurich: Christoph Froschauer.

Gessner, Konrad. 1548. *Pandectarum sive Partitionum Universalium*. Zurich: Christoph Froschauer.

Gessner, Konrad. 1549. *Partitiones Theologicae, Pandectarum Universalium*. Zurich: Christoph Froschauer.

Giebichenstein, Karl. 1929. Tempo... Das Maschinenprinzip der kürzesten Zeit. *Büro-Organisation* 3 (4):26–27.

Giesecke, Michael. 1998. *Der Buchdruck in der frühen Neuzeit: Eine historische Fallstudie über die Durchsetzung neuer Informations- und Kommunikationstechnologien*. Vol. 1357 of *stw*. Frankfurt am Main: Suhrkamp Verlag.

Gödel, Kurt. 1931. Über formal unentscheidbare Sätze der "Principia Mathematica" und verwandter Systeme I. *Monatshefte für Mathematik und Physik* 38:173–198.

Goethe, Johann Wolfgang von. 1801/1994. Tag- und Jahreshefte. In *Autobiographische Schriften II*. Vol. 10 of *Johann Wolfgang von Goethe: Werke: Hamburger Ausgabe*, 10, 429–528. Revised edition. Munich: C. H. Beck.

Goethe, Johann Wolfgang von. 1831/1996. *Faust: Der Tragödie Zweiter Teil. In fünf Akten*. Vol. 3 of *Johann Wolfgang von Goethe: Werke: Hamburger Ausgabe*, 16, 46–364. Revised edition. Munich: C. H. Beck.

Gosch, Josias Ludwig. 1789. *Fragmente über den Ideenumlauf*. Copenhagen: Proft.

Graesel, Arnim. 1902. *Handbuch der Bibliothekslehre: Zweite, voellig umgearbeitete Auflage der "Grundzüge der Bibliothekslehre, Neubearbeitung von Dr. Jul. Petzholdts Katechismus der Bibliothekslehre." Mit 125 Abbildungen und 22 Schrifttafeln*. Leipzig: Verlagsbuchhandlung von J. J. Weber.

Grafton, Anthony. 1990. *Forgers and Critics: Creativity and Duplicity in Western Scholarship*. Princeton, N.J.: Princeton University Press.

Grassauer, Ferdinand. 1883. *Handbuch für österreichische Universitäts- und Studien-Bibliotheken sowie für Volks-, Mittelschul- und Bezirks-Lehrerbibliotheken. Mit einer Sammlung von Gesetzen, amtlichen Entschliessungen, Verordnungen, Erlässen, Acten und Actenauszügen*. Vienna: Verlag von Carl Graeser.

Grimm, Jacob, and Wilhelm Grimm. 1854. *Deutsches Wörterbuch: Erster Band A—Biermolke*. Leipzig: Verlag von S. Hirzel.

Grimm, Jacob. 1956. *Deutsches Wörterbuch: Zwölfter Band I. Abteilung V—Verzwunzen*. Leipzig: Verlag von S. Hirzel.

Gumbrecht, Hans Ulrich. 1978. Modern, Modernität, Moderne. In *Geschichtliche Grundbegriffe: Historisches Lexikon zur politisch-sozialen Sprache in Deutschland*. Vol. 4., ed. Otto Brunner, Werner Conze, and Reinhart Koselleck, 92–131. Stuttgart: Klett-Cotta.

Gutkas, Karl. 1989. *Kaiser Joseph II: Eine Biographie*. Vienna: Paul Zsolnay Verlag.

Hagen, Wolfgang. 1993. Computerpolitik. In *Computer als Medium*, ed. Norbert Bolz, Friedrich Kittler, and Georg Christoph Tholen, 139–160. Vol. 4 of *Literatur- und Medienanalysen*. Munich: Wilhelm Fink Verlag.

Hagen, Wolfgang, ed. 204. *Warum haben Sie keinen Fernseher, Herr Luhmann? Letzte Gespräche mit Niklas Luhmann*. Berlin: Kulturverlag Kadmos.

Hainz-Sator, Werner. 1988. *Katalog der abendländischen Handschriften der Universitätsbibliothek Wien*. Vol. 146 of *Biblos-Schriften*. Vienna: Universitätsbibliothek.

References

Hapke, Thomas. 1997. Wilhelm Ostwald und seine Initiativen zur Organisation und Standardisierung naturwissenschaftlicher Publizistik: Enzyklopädismus, Internationalismus und Taylorismus am Beginn des 20. Jahrhunderts. In *Fachschrifttum, Bibliothek und Naturwissenschaft im 19. und 20. Jahrhundert*, ed. Christoph Meinel, 157–174. Vol. 27 of *Wolfenbütteler Schriften zur Geschichte des Buchwesens*. Wiesbaden: Otto Harrassowitz Verlag.

Harsdörffer, Georg Philipp. 1653/1990. *Delitiae Philosophicae et Mathematicae: Der Philosophischen und Mathematischen Erquickstunden Dritter Teil*. Vol. 3 of *Texte der Frühen Neuzeit*. Frankfurt am Main: Keip Verlag. Reprint of the 1653 Nürnberg edition, ed. and introduced by Jörg Jochen Berns.

Harsdörffer, Georg Philipp, and Daniel Schwenter. 1636/1991. *Deliciae Physico-Mathematicae oder Mathematische und Philosophische Erquickstunden Band 1*. Vol. 3 of *Texte der Frühen Neuzeit*. Frankfurt am Main: Keip Verlag. Reprint of the 1636 Nürnberg edition, ed. and introduced by Jörg Jochen Berns.

Haupt, Hermann. 1888. Eine Notiz über Kapseln zur Aufbewahrung des Blätterkatalogs. *Centralblatt für Bibliothekswesen* 5 (8):362–364.

Haußmann, Hermann. 1925. *Die Büroreform als Teil der Verwaltungsreform*. Berlin: Heymann.

Heidegger, Martin. 1954/1977. The Question Concerning Technology. In *The Question Concerning Technology and Other Essays*, 115–154. Trans. William Lovitt. New York: Harper & Row.

te Heesen, Anke. 2002. *The World in a Box: The Story of an Eighteenth-Century Picture Encyclopedia*. Chicago: University of Chicago Press.

te Heesen, Anke. 2005. The Notebook: A Paper Technology. In *Making Things Public*, ed. Bruno Latour and Peter Weibel, 263–286. Cambridge, Mass.: MIT Press.

te Heesen, Anke. 2006. *Der Zeitungsausschnitt: Ein Papierobjekt der Moderne*. Frankfurt am Main: Fischer Taschenbuch Verlag.

Heine, Heinrich. 1826/1995. *Reisebilder: Zweiter Teil. Ideen. Das Buch Le Grand*. Vol. 2. 3rd revised and expanded edition. Munich: Carl Hanser Verlag.

Hempel-Kürsinger, Johann-Nepomuk Freiherr von. 1825. *Alphabetisch-chronologische Übersicht der k.k. Gesetze und Verordnungen vom Jahre 1740 bis zum Jahre 1843 als Hauptrepertorium über die politischen Gesetzessammlungen*. Vol. 2. Vienna: Mösle's Wittwe.

Hempel-Kürsinger, Johann-Nepomuk Freiherr von. 1826. *Alphabetisch-chronologische Übersicht der k.k. Gesetze und Verordnungen vom Jahre 1740 bis zum Jahre 1843 als Hauptrepertorium über die politischen Gesetzessammlungen*. Vol. 6. Vienna: Mösle's Wittwe.

Herrmann, Elsa. 1931. *Kartei und Archiv.* Hamburg: Ansporn-Verlag Hans A. Blum.

Hildesheimer, Wolfgang. 1986. Die "Letzten Zettel." *Text & Kritik* 89/90:8–18.

Hilsenbeck, Adolf. 1912. Zur Frage einheitlicher Katalogisierungsregeln. *Centralblatt für Bibliothekswesen* 29 (7/8):310–321.

Hinnenthal, Hans. 1927. *Die deutsche Rationalisierungsbewegung und das Reichskuratorium für Wirtschaftlichkeit.* Berlin: Reichskuratorium für Wirtschaftlichkeit.

Hittmair, Anton. 1901. Zur Geschichte der österreichischen Bibliotheks-Instruktion. *Mitteilungen des Österreichischen Vereins für Bibliothekswesen,* 5, 9ff.

Hoffmeister, Johannes, ed. 1936. *Dokumente zu Hegels Entwicklung.* Stuttgart: Fr. Frommans Verlag.

Hofmann, Walter. 1916. *Buch und Volk und die volkstümliche Bücherei.* Vol. 4 of *Schriften der Zentralstelle für Volkstümliches Büchereiwesen.* Leipzig: Theodor Thomas Verlag.

Hofstadter, Douglas R. 1989. *Gödel, Escher, Bach: Ein endloses geflochtenes Band.* 12th edition. Stuttgart: Klett-Cotta.

Hohn, Thomas. 1993. Moderne Zeiten! Von der Karteikarte zur elektronischen Akte. *Jahrbuch der rheinischen Denkmalpflege* 36:517–524.

Holst, Helge. 1937. Blattkatalog oder Kartothek als systematischer Bibliothekskatalog. *Zentralblatt für Bibliothekswesen* 54 (11):556–564.

Hopkins, Judith. 1992. The 1791 French Cataloging Code and the Origins of the Card Catalog. *Libraries & Culture* 27 (4):378–404.

Hortzschansky, Adalbert. 1908. *Die Königliche Bibliothek zu Berlin: Ihre Geschichte und ihre Organisation. Vier Vorträge.* Berlin: Königliche Bibliothek.

Hottinger, Christlieb Gotthold. 1911. *Ein Bücher-Zettel-Katalog und ein Bio-Ikono-Bibliographisches Sammelwerk.* Berlin-Südende: Self-published.

Hummel, Otto. 1928. Zur Beweiskraft der Buchhaltung auf losen Blättern. *Organisation—Betrieb—Büro* 30 (7):160–162.

Ingold, Felix Philipp. 1993. Das Buch. In *Aisthesis: Wahrnehmung heute oder Perspektiven einer anderen Ästhetik,* ed. Karlheinz Barck, Peter Gente, Heidi Paris, and Stefan Richter, 289–294. Vol. 1352 of *Reclam-Bibliothek.* 5th revised edition. Leipzig: Reclam.

Institut International de Bibliographie. 1914. *Das Internationale Institut für Bibliographie in Brüssel: Organisation—Methoden—Sammlungen—Arbeiten—Veröffentlichungen—Kataloge.* Vol. 124 of *Publication.* Brussels: Internationales Institut für Bibliographie.

Ippel, Eduard. 1916. *Erinnerungen an die Königliche Bibliothek 1874–1885.* Freiburg im Breisgau: C. A. Wagners Hof- und Univ.-Buchdruckerei.

James, M. S. R. 1902. The Progress of the Modern Card Catalog Principle. *Public Libraries* 7:187.

Jean Paul. 1796a/1987. *Leben des Quintus Fixlein, aus funfzehn Zettelkästen gezogen; nebst einem Musteil und einigen Jus de tablette*. Stuttgart: Philipp Reclam jun.

Jean Paul. 1796b/1996. Die Taschenbibliothek. In *Sämtliche Werke*, part 2, vol. 3. *Jugendwerke und vermischte Schriften*, ed. Norbert Miller, 769–773. Vol. 3 of *Jean Paul—Sämtliche Werke*. Frankfurt am Main: Zweitausendeins.

Jean Paul. 1804/1996. *Vorschule der Ästhetik*. Vol. 5 of *Jean Paul—Sämtliche Werke*, ed. Norbert Miller, part 1. Munich: Carl Hanser Verlag.

Jesinger, Alois. 1926. *Kataloge und Aufstellung der Wiener Universitätsbibliothek in ihrer geschichtlichen Entwicklung*. Vol. 17 of *Bertholddruck*. Berlin: Berthold.

Jewett, Charles C. 1853. *Smithsonian Report on the Construction of Catalogues of Libraries and Their Publication by Means of Separate, Stereotyped Titles*. 2nd edition. Washington, D.C.: Smithsonian Institution.

Jochum, Uwe. 1991. *Bibliotheken und Bibliothekare 1800–1900*. Würzburg: Königshausen und Neumann.

Jochum, Uwe. 1993. *Kleine Bibliotheksgeschichte*. Stuttgart: Philipp Reclam jun.

Jochum, Uwe. 2003. Goethes Bibliotheksökonomie. In *Europa: Kultur der Sekretäre*, ed. Bernhard Siegert and Joseph Vogl, 111–123. Zurich: Diaphanes.

Junker, Carl. 1896. Über das zu wählende Format der Repertoriumszettel. *Bulletin de l'Institut international de bibliographie* 1:196–201.

Junker, Carl. 1897. *Das Internationale Institut für Bibliographie in Brüssel*. Vol. 19 of *Office International de Bibliographie*. Leipzig: Ramm & Seemann.

Kafka, Franz. 1925/1994. *Der Proceß: Roman in der Fassung der Handschrift*. Vol. 3 of *Gesammelte Werke in zwölf Bänden: Nach der kritischen Ausgabe herausgegeben von Hans-Gerd Koch*. Frankfurt am Main: Fischer Taschenbuch Verlag.

Kaiser, Rudolf. 1921. *Der alphabetische Zettelkatalog: Ein geschichtlicher Rückblick*. Berlin: Königliche und Staatsbibliothek.

Kant, Immanuel. [1797] 1996. *The Metaphysics of Morals*. Trans. Mary J. Gregor. Cambridge: Cambridge University Press.

Kayser, Albrecht Christoph. 1790. *Ueber die Manipulation bey der Einrichtung einer Bibliothek und der Verfertigung der Bücherverzeichnisse nebst einem alphabetischen Kataloge aller von Johann Jakob Moser einzeln herausgekommener Werke—mit Ausschluß seiner theologischen—und einem Register*. Bayreuth: Verlag der Zeitungsdruckerei.

Keysser, Adolf. 1885. Ueber die Einrichtung der alphabetischen Hauptkataloge öffentlicher Bibliotheken. *Centralblatt für Bibliothekswesen* 2 (1):1-10.

Kieser, Marco. 1997. Aus der Amtsbibliothek—Vom Zettel zur Datei. *Denkmalpflege im Rheinland* 14 (1):22-25.

Kittler, Friedrich. 1979. Vergessen. In *Texthermeneutik: Aktualität, Geschichte, Kritik*, ed. Ulrich Nassen, 195-221. Paderborn: Ferdinand Schöningh Verlag.

Kittler, Friedrich. 1986/1999. *Gramophone, Film, Typewriter*. Stanford, Calif.: Stanford University Press.

Kittler, Friedrich. 1988. Rhetorik der Macht und Macht der Rhetorik—Lohensteins Agrippina. In *Johann Christian Günther (mit einem Beitrag zu Lohensteins "Agrippina")*, ed. Hans-Georg Pott, 39-52. Paderborn: Ferdinand Schöningh.

Kittler, Friedrich. 1989. *Die Nacht der Substanz*. Bern: Benteli Verlag.

Kittler, Friedrich. 1993. Geschichte der Kommunikationsmedien. In *Raum und Verfahren*, ed. Aleida Assmann, Jörg Huber, and Alois Martin Müller, 169-188. Vol. 2 of *Interventionen*. Basel: Stroemfeld/Roter Stern.

Kittler, Friedrich. 1995. *Aufschreibesysteme 1800-1900*. 3rd completely revised edition. Munich: Wilhelm Fink Verlag.

Kittler, Friedrich. 1997. Memories Are Made of You. In *Schrift, Medien, Kognition: Über die Exteriorität des Geistes*, ed. Peter Koch and Sybille Krämer, 187-203. Vol. 19 of *Probleme der Semiotik*. Tübingen: Stauffenburg Verlag.

Kittler, Friedrich. 1998. Die Herrschaft der Schreibtische. In *Work@Culture: Büro. Inszenierung von Arbeit*, ed. Herbert Lachmeyer and Eleonora Louis, 39-42. Klagenfurt: Ritter Verlag.

von Kleist, Heinrich. 1805/1997. On the Gradual Production of Thoughts Whilst Speaking. In *Selected Writings*, ed. and trans. David Constantine, 405-409. Indianapolis, Ind.: Hackett Publishing.

Klinckowström, Carl Graf von. 1934. Die Kartei. *Zeitschrift für Bücherfreunde* 1:14-16.

Krajewski, Markus. 2001. Ver(b)rannt im *Fahlen Feuer*: Ein Karteikartenkommentar. *Kunstforum International* 155:288-292.

Krajewski, Markus, ed. 2004a. *Projektemacher: Zur Produktion von Wissen in der Vorform des Scheiterns*. Berlin: Kulturverlag Kadmos.

Krajewski, Markus. 2004b. Zum Glück: Die Organisation der Welt um 1900, *Neue Rundschau*, special issue. *Standards* 115 (4):39-59.

Krajewski, Markus. 2006. *Restlosigkeit: Weltprojekte um 1900*. Frankfurt am Main: Fischer Taschenbuch Verlag.

Krajewski, Markus. 2010. *Der Diener: Mediengeschichte einer Figur zwischen König und Klient.* Frankfurt am Main: S. Fischer Verlag.

Krajewski, Markus. 2011. Paper as Passion: Niklas Luhmann and His Card Index. In *Raw Data*, ed. Lisa Gitelman. Cambridge, Mass.: MIT Press.

Kunert, Günter. 1986. Zettel. *Akzente* 33 (5):391–394.

Küster, Hansjörg. 2000. Zinsen der Datenbankiers: Braucht die Gelehrtenrepublik ein Zitierkartellamt? *Frankfurter Allgemeine Zeitung* (August 7):45.

Lackmann, Heinrich. 1966. Leibniz' bibliothekarische Tätigkeit in Hannover. In *Leibniz: Sein Leben, sein Wirken, seine Welt*, ed. Wilhelm Totok and Carl Haase, 321–348. Hannover: Verlag für Literatur und Zeitgeschehen.

Ladewig, Paul. 1912. *Politik der Bücherei.* Leipzig: Ernst Wiegandt Verlagsbuchhandlung.

Ladewig, Paul. 1917. *Die öffentliche Bücherei.* Vol. 1 of *Schriften der Zentrale für Volksbücherei.* Berlin: Weidmann.

La Fontaine, Henri-Marie. 1903. Une mémoire mondiale. Le Répertoire bibliographique universel. *La Revue: Ancienne revue des revues*, 14, October 15, 201–208.

Lehnus, Donald J. 1974. *Milestones in Cataloging: Famous Catalogers and Their Writings, 1835–1969.* Littleton, Colo.: Libraries Unlimited.

Leibniz, Gottfried Wilhelm. 1679/1903. Consilium de Encyclopaedia nova conscribenda methodo inventoria. In *Opuscules et fragment inédits*, ed. Louis Couturat, 30–41. Paris.

Leibniz, Gottfried Wilhelm. [1696] 1987. *Allgemeiner politischer und historischer Briefwechsel*, ed. Gerda Utermöhlen. Darmstadt: Reichl.

Leibniz, Gottfried Wilhelm. 1699. Notwendige bibliothekarische Maßnahmen in der Bibliotheca Augusta. In *Braunschweigisches Jahrbuch* 54: 195.

von Leon, Gottlieb. 1820. *Kurzgefaßte Beschreibung der k.k. Hof-Bibliothek in Wien.* Vienna: Carl Armbruster's Verlag.

Leroi-Gourhan, André. 1993. *Gesture and Speech.* Cambridge, Mass.: MIT Press.

Lesky, Erna. 1973. Gerard van Swieten: Auftrag und Erfüllung. In *Gerard van Swieten und seine Zeit*, ed. Erna Lesky and Adam Wandruszka, 11–33. Vol. 8 of *Studien zur Geschichte der Universität Wien.* Vienna: Verlag Hermann Böhlaus Nachf.

Lessing, Gotthold Ephraim. 1748/1989. Der junge Gelehrte: Ein Lustspiel in drei Aufzügen. In *Gotthold Ephraim Lessing Werke 1743–1750*, ed. Jürgen Stenzel. Vol. 1 of *Gotthold Ephraim Lessing Werke und Briefe in zwölf Bänden*, ed. Wilfried Barner. Frankfurt am Main: Deutscher Klassiker Verlag.

Levy, Pierre. 1998. Die Erfindung des Computers. In *Elemente einer Geschichte der Wissenschaften*, ed. Michel Serres, 905–944. Frankfurt am Main: Suhrkamp Verlag.

Ley, Willy. 1929. *Konrad Gesner: Leben und Werk*. Vol. 15/16 of *Münchener Beiträge zur Geschichte und Literatur der Naturwissenschaften und Medizin*. Munich: Münchner Drucke.

Leyh, Georg. 1914. Systematische oder mechanische Aufstellung? *Zentralblatt für Bibliothekswesen* 31 (9/11):398–407.

Leyh, Georg. 1921. Chr. G. Heynes Eintritt in die Göttinger Bibliothek. In *Aufsätze Fritz Milkau gewidmet*, ed. Georg Leyh, 220–228. Leipzig: Verlag von Karl W. Hiersemann.

Leyh, Georg. 1929. *Das Büchermagazin in seiner Entwicklung*. Berlin: Elsnerdruck.

Leyh, Georg. 1961a. Aufstellung und Signaturen. In *Bibliotheksverwaltung*, 684–734. Vol. 2 of *Handbuch der Bibliothekswissenschaft. Begründet von Fritz Milkau*. 2nd expanded and revised edition. Wiesbaden: Otto Harrassowitz Verlag.

Leyh, Georg, ed. 1961b. *Bibliotheksverwaltung*. Vol. 2 of *Handbuch der Bibliothekswissenschaft. Begründet von Fritz Milkau*. 2nd expanded and enhanced edition. Wiesbaden: Otto Harrassowitz Verlag.

Le Coutre, Walter. 1931. Wesen, Wirtschaftlichkeit und Anwendungsgebiete von Karteien. *Büro-Organisation* 5 (7):54–55.

Library Bureau. 1902. *Library Catalog: A Descriptive List with Prices of the Various Articles of Furniture and Equipment for Libraries and Museums Furnished by the Library Bureau*. Boston: The Library Bureau.

Library Bureau. 1903. *Library Bureau System of Vertical Filing with Interchangeable Unit Cabinets*. Boston: Library Bureau.

Library Bureau. 1909. *The Story of Library Bureau*. Boston: Cowen Company, The University Press.

Library Bureau. n.d. *L.B. Armorclad Guides: Facts and Figures for the Sales Manager*. London: Library Bureau.

Library Bureau. n.d. *The Newest Force in Business Building: A Brief Description of the Five Big Divisions of Library Bureau Service*. London: Library Bureau.

Lindeboom, G. A. 1973. Gerard van Swietens erster Lebensabschnitt (1700–1745). In *Gerard van Swieten und seine Zeit*, ed. Erna Lesky and Adam Wandruszka, 63–79. Vol. 8 of *Studien zur Geschichte der Universität Wien*. Vienna: Verlag Hermann Böhlaus Nachf.

Löffler, Karl. 1956. *Einführung in die Katalogkunde*. 2nd newly revised edition. Stuttgart: Anton Hiersemann.

Löffler, Klemens. 1917/1918. Leibniz als Bibliothekar: Ein Nachtrag zum Leibnizjubiläum. *Zeitschrift für Bücherfreunde* 9 (4):95–100.

Long, Orie W. 1935. *Literary Pioneers: Early American Explorers of European Culture.* Cambridge, Mass.: Harvard University Press.

Lovett, Robert W. 1963. William Croswell, the Eccentric Scholar. Unpublished ms. Harvard University Archives, Cambridge, Mass.

Luhmann, Niklas. 1993. Kommunikation mit Zettelkästen: Ein Erfahrungsbericht. In *Universität als Milieu*, ed. André Kieserling, 53–61. Bielefeld: Haux.

Luhmann, Niklas. 1987. In *Archimedes und wir: Interviews*, ed. Dirk Baecker and Georg Stanitzek. Berlin: Merve Verlag.

May, Leo. 1997. *The So-Called Happiness Formula of Wilhelm Ostwald.* Vol. 463 of *Theorie und Forschung.* Regensburg: Roderer.

Meinel, Christoph. 1995. Enzyklopädie der Welt und Verzettelung des Wissens: Aporien der Empirie bei Joachim Jungius. In *Enzyklopädien der Frühen Neuzeit. Beiträge zu ihrer Erforschung*, ed. Franz M. Eybl, Wolfgang Harms, Hans-Henrik Krummacher, and Werner Welzig, 162–187. Vienna: Österreichische Akademie der Wissenschaften; Tübingen: Akademie der Wissenschaften und Literatur; Mainz: Max Niemeyer Verlag.

Menzel, M. 1926. *Bürokunde: Schulausgabe.* Vol. 1 of *Bücherei für Bürobetriebstechnik.* Berlin: Verlag für Bürotechnik.

Meyer, Richard M. 1907. "Vollständigkeit": Eine methodologische Skizze, *Euphorion. Zeitschrift für Literaturgeschichte* 14:1–17.

Meyers Großes Konversations=Lexikon. 1906. Article "Buchdruckerkunst." Vol. 3, 6th newly revised edition, 528–532. Leipzig: Bibliographisches Institut.

Meynen, Gloria. 1997. Bürokratien von Imperien, Bibliotheken und Maschinen. M.A. thesis, Humboldt Universität Berlin.

Meynen, Gloria. 1998. Büroformate: Von DIN A 4 zu Apollo 11. In *Work@Culture: Büro. Inszenierung von Arbeit*, ed. Herbert Lachmeyer and Eleonora Louis, 81–88. Klagenfurt: Ritter Verlag.

Michel, Paul. 2002. Ordnungen des Wissens: Darbietungsweisen des Materials in Enzyklopädien. In *Populäre Enzyklopädien: Von der Auswahl, Ordnung und Vermittlung des Wissens*, ed. Ingrid Tomkowiak, 35–83. Zurich: Chronos Verlag.

Milkau, Fritz. 1898. *Centralkataloge und Titeldrucke: Geschichtliche Erörterungen und praktische Vorschläge im Hinblick auf die Herstellung eines Gesamtkatalogs der preußischen wissenschaftlichen Bibliotheken.* Leipzig: Otto Harrassowitz.

Milkau, Fritz. 1912. Die Bibliotheken. In *Die allgemeinen Grundlagen der Kultur der Gegenwart*, part 1 of *Die Kultur der Gegenwart: Ihre Entwickelung und ihre Ziele*, ed. Paul Hinneberg, 580–629. 2nd revised and expanded edition. Berlin: G. Teubner.

Mitchell, Barbara A. 2003. "A Beginning Is Made": The New Card Catalog of the Harvard College Library and the Female Labor Force, 1856–1877. *Harvard Library Bulletin* 14 (Fall):11–32.

Molbech, Christian. 1833. *Über Bibliothekswissenschaft oder Einrichtung und Verwaltung öffentlicher Bibliotheken*. Leipzig: Verlag der J. C. Hinrichschen Buchhandlung.

Montaigne, Michel de. [1588] 2003. *The Complete Essays*. Trans. M.A. Screech. London: Penguin.

Morzé, Adolf von. 1982. Ladewig, Paul. In *Neue Deutsche Biographie*, ed. Historische Kommission bei der Bayrischen Akademie der Wissenschaften, 392–393. Vol. 13. Berlin: Duncker & Humblot.

Mosel, Ignaz F. Edlen von. 1835. *Geschichte der k.k. Hofbibliothek zu Wien*. Vienna: Fr. Beck'sche Universitäts-Buchhandlung.

Moser, Johann Jacob. 1773. *Vortheile vor Canzleyverwandte und Gelehrte in Absicht auf Akten-Verzeichnisse, Auszüge und Register, desgleichen auf Sammlungen zu künfftigen Schrifften und würckliche Ausarbeitung derer Schrifften*. Tübingen: Heerbrandt.

Moser, Johann Jacob. 1777. *Lebensgeschichte, von ihm selbst beschrieben. Dritter und letzter Theil*. 3rd edition. Frankfurt: n.p.

Mühlpfordt, Wolfgang. 1908. Die gesetzliche Buchführungspflicht und das Dauer-Konto-"Buch." *System: Zeitschrift für moderne Geschäfts- und Betriebskunde* 1:100–102.

Müller, Götz. 1988. *Jean Pauls Exzerpte*. Würzburg: Königshausen & Neumann.

Müller, Heiner. 1977/1978. Hamletmaschine. In *Mauser*, 89–97. Berlin: Rotbuch/Verlag der Autoren.

Müller, Heiner. 1986/1994. Wolokolamsker Chaussee IV Kentauren. In *Shakespeare Factory 2*, 245–250. Vol. 9 of *Heiner Müller: Texte*. Berlin: Rotbuch Verlag.

Müller, Heiner. 1995. Mommsens Block. In *Die Unschreibbarkeit von Imperien: Theodor Mommsens Römische Kaisergeschichte und Heiner Müllers Echo*, ed. Wolfgang Ernst, 41–47. Weimar: Verlag für Datenbank und Geisteswissenschaften.

Murr, Christoph Gottlieb von. 1779. Von Leibnitzens Exzerpirschrank. *Journal zur Kunstgeschichte und allgemeinen Litteratur* 7:210–212.

Musil, Robert. [1932] 1995. *The Man without Qualities*. Trans. Sophie Wilkinson. New York: Alfred A. Knopf.

Naudé, Gabriel. [1627] 1963. *Advis pour dresser une bibliothèque*. Leipzig: VEB Edition.

Negelinus, Johannes, and Mox Doctor. 1913. *Schattenrisse*. Leipzig: Skiamacheten Verlag.

Neuland, Eva. 1995. Stand und Entwicklung der Kartei arbeitsloser und von Arbeitslosigkeit bedrohter habilitierter Germanistinnen und Germanisten. *Mitteilungen des deutschen Germanistenverbandes* 42 (1):77–79.

Niemann, Willi Bruno. 1927. *Das Dewey-System (Dezimal-Klassifikation) und seine Verwendbarkeit für Bibliotheken und Literatur-Karteien*. Vol. 2 of *Wege zu technischen Büchern*. Berlin: Verlag von Robert Kiepert.

Osterwald, Albert. 1914. Geschichte und Entwicklung der Schreibmaschine. *Archiv für Buchgewerbe* 51:252–254.

Ostwald, Wilhelm. 1912a. *Der energetische Imperativ*. Leipzig: Akademische Verlagsgesellschaft.

Ostwald, Wilhelm. 1912b. *Die Organisierung der Organisatoren durch die Brücke*. Munich: Selbstverlag der Brücke.

Ostwald, Wilhelm. 1912c. *Sekundäre Weltformate*. Ansbach: Fr. Seybold's Buchhandlung.

Ostwald, Wilhelm. 1914. Referat über den "VI. Kongress des internationalen Verbandes für die Materialprüfungen der Technik." *Zeitschrift für Physikalische Chemie* (87):119.

Ostwald, Wilhelm. 1929. Organisation der Organisation. *Zeitschrift für Organisation* 3 (9):229–232.

Ostwald, Wilhelm. 1933. *Lebenslinien: Eine Selbstbiographie*. Berlin: Verlag Klasing & Co.

Paulukat, A. 1910. Lose Blätter in der Buchführung. *Bürobedarfs-Rundschau* 8:513–514.

Perrot, Franz. 1874. *Das Bankwesen und die Zettel-Privilegien: Eine Untersuchung über die Principien der modernen Wirtschaftsführung und über das Verhältnis des Bank-Gewerbes zu denselben*. Rostock: Stiller'sche Hof- und Universitätsbuchhandlung.

Petschar, Hans. 1993. *Niederländer, Europäer, Österreicher: Hugo Blotius, Sebastian Tengnagel, Gérard Freiherr van Swieten, Gottfried Freiherr van Swieten: Vier Präfekten der kaiserlichen Hofbibliothek in Wien*. Vol. 1993, 3 of *Sonderausstellungen*. Vienna: Österreichische Nationalbibliothek.

Petschar, Hans. 1999. Einige Bemerkungen, die sorgfältige Verfertigung eines Bibliothekskatalogs für das allgemeine Lesepublikum betreffend. In *Der Zettelkatalog: Ein historisches System geistiger Ordnung*, ed. Hans Petschar, Ernst Strouhal, and Heimo Zobernig, 17–43. Vienna: Springer Verlag.

Petschar, Hans, Ernst Strouhal, and Heimo Zobernig, eds. 1999. *Der Zettelkatalog: Ein historisches System geistiger Ordnung*. Vienna: Springer Verlag.

Pfeifer, Wolfgang, ed. 1995. *Etymologisches Wörterbuch des Deutschen*. Vol. 3358. 2nd edition. Munich: dtv.

Pick, Albert. 1967. *Papiergeld: Ein Handbuch für Sammler und Liebhaber*. Braunschweig: Klinkhardt & Biermann.

Pinner, Felix. 1918. *Emil Rathenau und das elektrische Zeitalter*. Vol. 6 of *Grosse Männer: Studien zur Biologie des Genies*, ed. Wilhelm Ostwald. Leipzig: Akademische Verlagsgesellschaft.

Placcius, Vincentius. 1689. *De Arte Excerpendi: Vom gelahrten Buchhalten*. Stockholm, Hamburg: Bei Gottfried Liebezeit.

Popp, Ewald. 1932. *Vom Arbeiten nach dem Prinzip der wissenschaftlichen Betriebsführung (Taylor-Prinzip): Erfahrungen und Gedanken aus der Praxis*. Köln: Franz Greven Verlag.

Porstmann, Walter. 1928a. *Karteikunde: Das Handbuch für Karteitechnik*. 2nd edition. Stuttgart: Verlag für Wirtschaft und Verkehr.

Porstmann, Walter. 1928b. "Kartei" und Amtsschimmel. *Organisation—Betrieb—Büro* 30 (15):359–361.

Porstmann, Walter. 1929. Kartei—Blattei—Mappei. *Zeitschrift für Organisation* 3:609–611.

Porstmann, Walter. 1939. *Karteikunde: Das Handbuch für Karteitechnik*. 3rd revised edition. Berlin: Spaeth & Linde.

Potter, Alfred Claghorn, and Charles Knowles Bolton. 1897. *The Librarians of Harvard College 1667–1877*. Vol. 52 of *Bibliographical Contributions*. Cambridge, Mass.: Library of Harvard University.

Predeek, Albert. 1927. Die Bibliotheken und die Technik. *Zentralblatt für Bibliothekswesen* 44 (9/10):462–485.

Rabinbach, Anson. 1992. *The Human Motor: Energy, Fatigue, and the Origins of Modernity*. Berkeley: University of California Press.

Rabofsky, Aladár. 1932. Zur Psychotechnik der Kartei. *Industrielle Psychotechnik* 9 (11/12):321–343.

Radlecker, Kurt. 1950. Gottfried van Swieten: Eine Biographie. Doctoral dissertation, phil., Vienna.

Rautenstrauch, Franz Stephan. 1778. Instruction vorgeschrieben für alle Universitäts- und Lycealbibliotheken mit Hofdecrete vom 30. April 1778, Z. 628. In *Handbuch für österreichische Universitäts- und Studien-Bibliotheken sowie für Volks-,*

Mittelschul- und Bezirks-Lehrerbibliotheken. Mit einer Sammlung von Gesetzen, a. Entschlissungen, Verordnungen, Erlässen, Acten und Actenauszügen, ed. Ferdinand Grassauer, 171–175. Vienna: Verlag von Carl Graeser.

Remington Arms Company. 1999. *History of the Firearms Business 1816–1998*. World Wide Web, http://www.remington.com.

Remington Rand GmbH. 1957. *UNIVAC-Informationen: Eine Sammlung praktischer Anwendungsbeispiele*. Frankfurt am Main: Remington Rand GmbH.

Reuleaux, Franz. 1875. *Theoretische Kinematik: Grundzüge einer Theorie des Maschinenwesens*. Vol. 1 of *Lehrbuch der Kinematik*. Braunschweig: Vieweg.

Riberette, Pierre. 1970. *Les bibliothèques françaises pendant la Révolution (1789–1795): Recherches sur un essai de catalogue collectif*. Vol. 2 of *Mémoires de la section d'histoire moderne et contemporaine*. Paris: Bibliothèque Nationale.

Roloff, Heinrich. 1961. Die Katalogisierung. In *Bibliotheksverwaltung*, ed. Georg Leyh, 242–356. Vol. 2 of *Handbuch der Bibliothekswissenschaft. Begründet von Fritz Milkau*. 2nd expanded and revised edition. Wiesbaden: Otto Harrassowitz Verlag.

Roloff, Heinrich. 1967. Die Renaissance des Bandkataloges. *Zentralblatt für Bibliothekswesen* 81 (5):267–273.

Rosenkranz, Karl. [1844] 1969. *Georg Wilhelm Friedrich Hegels Leben*, 2nd unmodified reprographic reprint of the Berlin, 1844 edition. Darmstadt: Wissenschaftliche Buchgesellschaft.

Rotella, Elyce J. 1981. The Transformation of the American Office: Changes in Employment and Technology. *Journal of Economic History* 41 (March):51–57.

Roth-Seefrid, Carl Friedrich. 1918. *Die Geisteskartothek: Ein zweckmäßiges Hilfsmittel im Kampf um unsere wirtschaftliche Existenz*. Munich: G. Franz'sche Hofbuchhandlung.

Rozier, Abbé François. 1775. *Nouvelle table des articles contenus dans les volumes de l'Académie Royale des Sciences de Paris depuis 1666 jusqu'en 1770*. Paris: Ruault.

Ruetz, Michael. 1993. *Arno Schmidt*. Frankfurt am Main: Zweitausendeins.

Saager, Adolf. 1911. *Die Brücke als Organisierungsinstitut*. Ansbach: Fr. Seybold's Buchhandlung.

Saager, Adolf. 1921. Die Brücke: Historisches. Unpublished typescript, archives of the Wilhelm-Ostwald-Forschungs- und Gedenkstätte. Großbothen.

Sachsse, Rolf. 2004. Das Gehirn der Welt: 1912. Die Organisation der Organisatoren durch die Brücke. In *Wilhelm Ostwald: Farbsysteme / Das Gehirn der Welt*, ed. Peter Weibel and Rolf Sachsse, 64–88. Ostfildern: Hatje Cantz Verlag.

Schaukal, Richard. 1913. *Zettelkasten eines Zeitgenossen: Aus Hans Bürgers Papieren*. Munich: Georg Müller.

Schaukal, Richard. 1918. *Erlebte Gedanken: Neuer Zettelkasten.* Munich: Georg Müller.

Scheel, Günter. 1973. Leibniz' Beziehungen zur Bibliotheca Augusta in Wolfenbüttel (1678–1716). *Braunschweigisches Jahrbuch* 54:172–199.

Scheffler, Karl. 1928. Neuzeitliche Büroorganisation. *Büro-Organisation* 2 (5):35.

Schellenberg, Theodore R. 1961. *Akten- und Archivwesen in der Gegenwart: Theorie und Praxis.* Munich: Karl Zink Verlag.

Schlesinger, Georg. 1920. *Psychotechnik und Betriebswirtschaft.* Vol. 1 of *Psychotechnische Bibliothek.* Leipzig: Hirzel.

Schmeizel, Martin. 1728. *Versuch zu einer Historie der Gelehrheit,* Jena: Zu finden bei Peter Fickelscherrn.

Schmidt, Arno. 1995. Der Platz, an dem ich schreibe. In *Essays und Aufsätze 2,* 28–31. Vol. 3, 4 of *Bargfelder Ausgabe.* Zurich: Haffmans Verlag.

Schmidt, Erich H. 1939. *Arbeitsplatzgestaltung im Büro.* Berlin: Deutscher Betriebswirte-Verlag.

Schmidt, Friedrich. 1922. *Die Pinakes des Kallimachos.* Vol. 1 of *Klassisch-Philologische Studien.* Berlin: Verlag von Emil Ebering.

Schmidt, Harald, and Marcus Sandl, eds. 2002. *Gedächtnis und Zirkulation: Der Diskurs des Kreislaufs im 18. und frühen 19. Jahrhundert.* Vol. 14 of *Formen der Erinnerung.* Göttingen: Vandenhoeck & Ruprecht.

Schmidt-Künsemüller, Friedrich Adolf. 1972. Gutenbergs Schritt in die Technik. In *Der gegenwärtige Stand der Gutenberg-Forschung,* ed. Hans Widmann, 122–147. Vol. 1 of *Bibliothek des Buchwesens.* Stuttgart: Anton Hiersemann Verlag.

Schneider, Birgit. 2007. *Textiles Prozessieren: Eine Mediengeschichte der Lochkartenweberei.* Zurich: Diaphanes Verlag.

Schneider, Manfred. 1994. *Liebe und Betrug: Die Sprache des Verlangens.* Munich: Deutscher Taschenbuch Verlag.

Schön, Erich. 1987. *Verlust der Sinnlichkeit oder die Verwandlungen des Lesers: Mentalitätswandel um 1800.* Vol. 12 of *Sprache und Geschichte.* Stuttgart: Klett-Cotta.

Schopenhauer, Arthur. [1819] 1966. *The World as Will and Representation.* Vol. 2. Trans. E. F. Payne. New York: Dover Publications.

Schramm, Dr. 1914. Erste deutsche Kriegsausstellung. *Archiv für Buchgewerbe* 51:277–280.

Schreiber, Heinrich. 1927. Quellen und Beobachtungen zur mittelalterlichen Katalogisierungspraxis besonders in deutschen Kartausen. *Zentralblatt für Bibliothekswesen* 44 (1/2):1–19.

References

Schreiber, Jens. 1983. *Das Sympton des Schreibens: Roman und absolutes Buch in der Frühromantik (Novalis/Schlegel)*. Vol. 649 of *Europäische Hochschulschriften*. Frankfurt am Main: Peter Lang Verlag.

Schrettinger, Martin. 1808. *Versuch eines vollständigen Lehrbuches der Bibliothek-Wissenschaft oder Anleitung zur vollkommenen Geschäftsführung eines Bibliothekars*. Vol. I. Munich: Selbstverlag.

Schulte, von. 1888. Rautenstrauch, Franz Stephan. In *Allgemeine Deutsche Biographie*, ed. Historische Commission bei der königl. Akademie der Wissenschaften, 459–460. Vol. 27. Leipzig: Duncker & Humblot.

Schunke, Ilse. 1927. Die systematischen Ordnungen und ihre Entwicklung: Versuch einer geschichtlichen Übersicht. *Zentralblatt für Bibliothekswesen* 44:377–400.

Scott, Edith. 1976. The Evolution of Bibliographic Systems in the United States, 1876–1945. *Library Trends* 25 (1):293–309.

Seeck, Otto. 1924. Laterculum. In *Paulys Realencyclopädie der classischen Altertumswissenschaft*, ed. Georg Wissowa. Vol. 12, col. 904–907. Stuttgart: Metzler.

Shannon, Claude E. [1948] 1993. A Mathematical Theory of Communication. In *Claude Elwood Shannon: Collected Papers*, ed. N. J. A. Sloane and Aaron D. Wyner, 5–83. New York: IEEE Press.

Sherman, E. W. 1916. History and Growth of Indexing Department. *L.B. Monthly News* 24:42–44. Supplement: 40th Anniversary Number 1876–1916.

Sibley, John Langdon. 1863. Librarian's Report, 12 July, 1861. In *Report of the Committee of the Overseers of Harvard College Appointed to Visit the Library for the Year 1862: Together with the Accompanying Documents*, ed. Harvard College Library, 14–35. Boston: Press of Geo C. Rand & Avery.

Siegert, Bernhard. 1993. *Relays: Literature as an Epoch of the Postal System*. Trans. Kevin Repp. Stanford, Calif.: Stanford University Press.

Smith, Minna C. 1902. Reorganizing Industries: A Novel Profession. *The World's Work* (5):2872.

Soennecken, Friedrich. 1914. Papierwaren, Schreibwesen, Malfarben. In *Amtlicher Katalog*, ed. Internationale Ausstellung für Buchgewerbe und Graphik, 117–122. Leipzig.

Staderini, Aristide. 1896. Zettel-Kataloge: Beschreibung und Preisverzeichnis. Via Dell'Archetto. 18, 19., Rome [n.d., after 1896].

Stadermann, Hans-Joachim. 1994. *Die Fesselung des Midas: Eine Untersuchung über den Aufstieg und Verfall der Zentralbankkunst*. Tübingen: J. C. B. Mohr (Paul Siebeck).

Steierwald, Ulrike. 1995. *Wissen und System: Zu Gottfried Wilhelm Leibniz' Theorie einer Universalbibliothek*. Vol. 22 of *Kölner Arbeiten zum Bibliotheks- und Dokumentationswesen*. Cologne: Greven Verlag.

Stieg, Margaret F. 1986. The Richtungstreit: The Philosophy of Public Librarianship in Germany before 1933. *Journal of Library History* 21 (2):261–276.

Stockhammer, Robert. 2000. Zwischenspeicher: Zur Ordnung der Bücher um 1800. In *Das Laokoon-Paradigma: Zeichenregime im 18. Jahrhundert*, ed. Michael Franz and Wolfgang Schäffner, 45–63. Berlin: Akademie Verlag.

Stoeckel, Herbert J. 1946. Loose-Leaf and Accounting. *Accounting Review* 21 (4):380–385.

Strouhal, Ernst. 1999. Zettel, Kasten, Katalog. In *Der Zettelkatalog: Ein historisches System geistiger Ordnung*, ed. Hans Petschar, Ernst Strouhal, and Heimo Zobernig, 9–16. Vienna: Springer Verlag.

Stummvoll, Josef, ed. 1968. *Geschichte der österreichischen Nationalbibliothek: Erster Teil: Die Hofbibliothek (1368–1922)*. Vol. 3 of *Museion. Veröffentlichungen der österreichischen Nationalbibliothek*. Vienna: Georg Prachner Verlag.

van Swieten, Gottfried. 1780. Vorschrift worauf die Abschreibung aller Bücher der k.k. Hofbibliothek gemacht werden solle. Unpublished ms., Austrian National Library, paper manuscript collection. Akt HB 125, 1 Teil.

van Swieten, Gottfried. 1787/1968. Bericht des Freiherrn Gottfried van Swieten über die Entwicklung der Hofbibliothek in den Jahren 1765–1787. In *Geschichte der österreichischen Nationalbibliothek. Erster Teil. Die Hofbibliothek (1368–1922)*, ed. Josef Stummvoll, 317–322. Vol. 3 of *Museion. Veröffentlichungen der österreichischen Nationalbibliothek*. Vienna: Georg Prachner Verlag.

Swift, Jonathan. [1728] 1992. *Gulliver's Travels*. Hertfordshire: Wordsworth Editions.

Tantner, Anton. 2007a. *Ordnung der Häuser, Beschreibung der Seelen: Hausnummerierung und Seelenkonskription in der Habsburgermonarchie*. Vol. 4 of *Wiener Schriften zur Geschichte der Neuzeit*. Innsbruck: Studien-Verlag.

Tantner, Anton. 2007b. *Die Hausnummer: Eine Geschichte von Ordnung und Unordnung*. Marburg: Jonas Verlag.

Taylor, Frederick Winslow. 1913. *Die Grundsätze wissenschaftlicher Betriebsführung*. Deutsche autorisierte Ausgabe von Rudolf Roesler. Munich: R. Oldenbourg Verlag.

Tenner, Edward. 1990. From Slip to Chip: How Evoking Techniques of Information Storage and Retrieval Have Shaped the Way We Do Mental Work. *Harvard Magazine* (11/12):52–57.

Thron, Josef. 1904. *Das Institut international de bibliographie in Brüssel: Gegenwärtiger Stand seiner Arbeiten und Veröffentlichungen.* Vol. 59a of *Publications de l'Office international de bibliographie, Institut international de bibliographie.* Brussels.

Ticknor, Anna Eliot, ed. 1874. *Life of Joseph Green Cogswell as Sketched in His Letters.* Cambridge, Mass.: Privately Printed at the Riverside Press.

Turing, Alan M. 1987. *Intelligence Service. Schriften.* Ed. Bernhard Dotzler and Friedrich Kittler. Berlin: Brinkmann & Bose.

Turner, Harold M. 1978. Cogswell, Joseph Green (1786–1871). In *Dictionary of American Library Biography,* ed. George S. Bobinski, Jesse H. Shera, and Bohdan S. Wynar, 87–91. Littleton, Colo.: Libraries Unlimited.

Unisys Corporation. 1999. *A History of Excellence.* World Wide Web, http://www.unisys.com.

Unterkircher, Franz. 1968. Hugo Blotius und seine ersten Nachfolger (1575–1663). In *Geschichte der österreichischen Nationalbibliothek: Erster Teil: Die Hofbibliothek (1368–1922),* ed. Josef Stummvoll, 79–162. Vol. 3 von *Museion. Veröffentlichungen der österreichischen Nationalbibliothek.* Vienna: Georg Prachner Verlag.

Utley, George Burwell. 1926. *Fifty Years of the American Library Association.* Chicago: American Library Association.

Utley, George Burwell. 1951. *The Librarians' Conference of 1853: A Chapter in American Library History.* American Library Association, Chicago.

Vann, Sarah K. 1978. *Melvil Dewey: His Enduring Presence in Librarianship.* Vol. 4 of *The Heritage of Librarianship Series.* Littleton, Colo.: Libraries Unlimited.

Vismann, Cornelia. [2000] 2008. *Files: Mediatechnique and the Law.* Trans. Geoffrey Winthrop-Young. Stanford, Calif.: Stanford University Press.

Vogel, Walter. 1931. Optische Orientierung durch Sichtkarteien! *Büro-Organisation* 5 (8):61.

Vogt, Victor. 1922. *Die Kartei: Ihre Anlage und Führung.* Vol. 5 of *Orga-Schriften.* 2nd edition, revised by Dr. Porstmann. Berlin: Organisation Verlagsanstalt.

Walton, Clarence E. 1939. *The Three-Hundredth Anniversary of the Harvard College Library.* Cambridge, Mass.: Harvard College Library.

Wangermann, Ernst. 1978. *Aufklärung und staatsbürgerliche Erziehung: Gottfried van Swieten als Reformator des österreichischen Unterrichtswesens 1781–1791.* Munich: R. Oldenbourg Verlag.

Warneyer, Otto, and Fritz Koppe. 1927. *Das Handelsgesetzbuch (ohne Seerecht) in der seit dem 22. April 1927 geltenden Fassung.* 2nd revised edition. Berlin: Industrieverlag Spaeth & Linde.

Wegmann, Nikolaus. 2000. *Bücherlabyrinthe: Suchen und Finden im alexandrinischen Zeitalter*. Cologne: Böhlau Verlag.

Weise, Franz. 1928. Kartenregister. *Büro-Organisation* 2 (9):65–66.

Wellisch, Hans H. 1981. How to Make an Index—16th Century Style: Conrad Gessner on Indexes and Catalogs. *International Classification* 8 (1):10–15.

Werle, Dirk. 2010. Die Bücherflut in der frühen Neuzeit—realweltliches Problem oder stereotypes Vorstellungsmuster? In *Frühneuzeitliche Stereotype: Zur Produktivität und Restriktivität sozialer Vorstellungsmuster*, ed. Miroslawa Czarnecka, et al., 469–486. Bern: Peter Lang.

Werner, Richard Maria. 1888. *Aus dem Josephinischen Wien: Geblers und Nicolais Briefwechsel während der Jahre 1771–1786*. Berlin: Verlag von Wilhelm Hertz.

Wiedemann, Conrad. 1967. Polyhistors Glück und Ende von Daniel Georg Morhof zum jungen Lessing. In *Festschrift Gottfried Weber: Zu seinem 70. Geburtstag überreicht von Frankfurter Kollegen und Schülern*, ed. Heinz Otto Burger and Klaus von See, 215–235. Vol. 1 of *Frankfurter Beiträge zur Germanistik*. Bad Homburg: Gehlen Verlag.

Wiegand, Wayne A. 1996. *Irrepressible reformer: A biography of Melvil Dewey*. Chicago: American Library Association.

Wieser, Walter G. 1968. Die Hofbibliothek in der Epoche der beiden van Swieten (1739–1803). In *Geschichte der österreichischen Nationalbibliothek. Part 1: Die Hofbibliothek (1368–1922)*, ed. Josef Stummvoll, 219–323. Vol. 3 of *Museion. Veröffentlichungen der österreichischen Nationalbibliothek*. Vienna: Georg Prachner Verlag.

Winchester, Simon. 2003. *The Meaning of Everything: The Story of the Oxford English Dictionary*. Oxford: Oxford University Press.

Winter, Eduard. 1943. *Der Josefinismus und seine Geschichte: Beiträge zur Geistesgeschichte Österreichs 1740–1848*. Vol. 1 of *Prager Studien und Dokumente zur Geistes- und Gesinnungsgeschichte Ostmitteleuropas*. Brunn: Rudolf M. Rohrer Verlag.

Witte, Irene Margarete. 1925. *Taylor, Gilbreth, Ford: Gegenwartsfragen der amerikanischen und europäischen Arbeitswissenschaft*. 2nd edition. Munich: Oldenbourg Verlag.

Witte, Irene Margarete. 1926. *Amerikanische Büroorganisation*. 2nd edition. Munich: Oldenbourg Verlag.

Witte, Irene Margarete. 1930. Amerikanische Büro-Organisation. In *Psychotechnik der Organisation in Fertigung, (Büro-)Verwaltung, Werbung*, ed. Johannes Wiedenmüller and Hans Piorkowski, 55–88. Vol. 5, 2 of *Handbuch der Arbeitswissenschaft. Objektspsychotechnik*. Halle: Marhold.

Wohlrab, Hertha, and Felix Czeike. 1972. Die Wiener Häusernummern und Straßentafeln: Ein Beitrag zu ihrer historischen Entwicklung. *Wiener Geschichtsblätter* 27 (2):333–351.

References

Wootton, Charles W., and Carel M. Wolk. 2000. The Evolution and Acceptance of the Loose-Leaf-Accounting System, 1885–1935. *Technology and Culture* 41 (1):80–98.

von Wurzbach, Constant. 1873. Rautenstrauch. In *Biographisches Lexikon des Kaisertums Österreich, enthaltend die Lebensskizzen der denkwürdigen Personen, welche seit 1750 in den österreichischen Kronländern geboren wurden oder darin gelebt und gewirkt haben*, 67–69. Vol. 25. Vienna: Druck und Verlag der k.k. Hof- und Staatsdruckerei.

Yates, JoAnne. 1989. *Control through Communication: The Rise of System in American Management*. Studies in Industry and Society. Baltimore, Md.: The Johns Hopkins University Press.

Zedelmaier, Helmut. 1992. *Bibliotheca universalis und bibliotheca selecta: Das Problem der Ordnung des gelehrten Wissens in der frühen Neuzeit*. Vol. 33 of *Beihefte zum Archiv für Kulturgeschichte*. Köln, Weimar, Wien: Böhlau.

Index

Abbot, Ezra, 78, 80, 82, 83, 91, 103, 107, 111, 165, 169
Abbot, John Lovejoy, 71
Access, 4, 28, 31, 45, 108, 112, 128–129
 to books, 29–31, 108, 151
 directed, 33, 51, 80, 88, 151
 public, 82, 103
 quick, 15, 127–128
 random, 32–33
Address, 30–31, 52, 121, 151
 of a book, 29, 50
 collection, 117–118
 of a house, 27–28, 31
 of an index card, 64
 numeric, 88
Addressing, 30–32, 156
 of books, 40
 exactness, 28
 of individuals, 28, 58
 of operations, 132
 postal, 29
Advertising stamps, 122
Algorithm13, 32, 40, 45–46, 70, 91, 120, 155
American Library Association, 91, 100, 166
 colonization, 107, 113
 founding, 87, 89
 Library Journal, 95
 second conference, 92
 Supply Department, 92
Ansbach, 122

Babbage, Charles, 175
Bach, Wilhelm H., 133, 135
Baker, Nicholson, 182
Banker's note, 58
Bank note, 7, 58–61, 160
Bankruptcy, intellectual, 66
Baroque rhetoric, 55, 149
Bartsch, Adam, 39–41, 44, 155
Benjamin, Walter, 135, 138
Berend, Eduard, 158
Bertillon, Alphons, 101, 125
Bible, 9, 42, 158, 160
Bibliography, 10, 42, 70, 80
 modern, 9–10
 national, 45–47
 objectives of, 16
 universal, 113–114
Blotius, Hugo, 16–17, 21, 35, 44, 148, 153–154
Boerhaave, Herrmann, 153
Book, 9, 52
 access to (*see* Access, to books)
 bound, 32, 41, 95, 170
 censorship, 36, 38, 154
 cover page, 151
 culture, 132
 description of a, 10, 44

Book (*cont.*)
  dimensions of a, 137–138
  disappearance, 139
  efficiency, 136
  excerpt, 17, 149
  flood (*see* Stream, of books)
  vs. index, 5, 54–55, 107, 127–128, 136–138, 169
  limits of the, 138
  locating a, 30
  loose-leaf, 126, 132
  loss, 153
  as medium, 8, 53, 55, 57, 135–136, 177
  printing, 9, 14
  production, 35
  as storage, 20
  as vehicle of civilization, 5, 128, 140
Book flood. *See* Stream, of books
Bookkeeping, 136, 138–140
  automated, 126, 139
  based on paper slips, 95, 131
  law on, 131
  Melvil Dewey's, 90, 94, 166
Bookshelf, 29, 33, 55, 108–109
Borchert, Jürgen, 161
Bound catalog. *See* Catalog
Box. *See* Card index
Bridge, The, 114, 120–121, 124, 126, 135, 140, 172–173
Brown, James Duff, 102
Brunet, Jacques Charles, 72, 162
Bücherei. *See* Library, public
BUGRA, 126–127
Bührer, Karl Wilhelm, 116–117, 119–122, 165, 172–174
Burchard, Gustav, 92
Burchardt, Lothar, 175
Burroughs, William S., 106
Business card, 89–90
Businessman, 6, 126, 133, 177–179
Buzás, Ladislaus, 153

Callimachus, 6, 145
Capital
  accumulating cultural, 62
  culture, 61
  flow, 59, 61
  representation of, 59–60
Carbon paper, 101, 226
Card, 47, 80
  business, 89–90
  format, 33
  playing, 33, 39, 47, 56, 89, 107, 152–153, 159
  post, 92
  punch, 6, 104, 111–112, 139, 142
Card catalog, 3, 5, 35, 42, 67, 70, 96, 111–112, 156
  appearance, 43, 91, 100, 106
  advantages, 78–79, 99, 107, 111
  Berlin, 108
  for bookkeeping, 96
  capsule, 41
  first, 4, 38
  Harvard, 70
  Hottinger's, 118
  hybrid, 75
  implementation, 42, 44–45, 47
  inventor of the, 102
  Josephinian, 31, 35, 37, 41–44, 91
  transformation into index, 95
Card index, 1, 5–8, 13, 15–16, 20, 67–68, 83, 111, 127, 138
  access to the, 128–129
  automated, 120
  vs. book, 129, 138, 164
  cabinet, 20
  capabilities, 108, 127–128, 130
  dimensions of the, 137–138
  enforcement of, 67–68, 124, 139–140
  etymology, 135, 147
  fingerprint, 101
  functions of, 50
  genealogy of, 1, 3, 70, 134, 139, 178
  of Hegel, 57, 159

Index

history of, 3, 5, 57, 139–140
hybrid, 13, 164
imagery, 179
of Jean Paul, 54
of J. J. Moser, 53–54
vs. *Kartothek*, 135
leader, 128
vs. library, 135
literary, 158, 161
machine, 1, 6, 128, 133, 136
as medium, 121, 136
patent protection, 133
protection of word, 133–134
scholarly, 13, 50, 67, 139
standardized, 53
system, 128–129
as vehicle of civilization, 128, 140
Card register. *See* Card index
Case. *See* Card index
Catalog, 16, 31, 38, 50, 87
  access to, 31, 111 (*see also* Access)
  author, 21
  basic, 44–45, 156
  card (*see* Index card)
  cutting up of, 163
  enforcement of, 32, 52, 151
  etymology of, 157
  by Ezra Abbot, 82, 169
  by Franz Rautenstrauch, 44, 156
  by Gottfried van Swieten, 34–35
  by G. W. Leibniz, 149–150
  Harvard College Library, 70–75, 77–79, 82–83
  by Hugo Blotius, 16
  library, 17, 32, 42–45, 47, 79, 81–82, 107, 111, 140, 155–156
  vs. local memory, 152
  of mud, 104
  opening of the, 80–81, 111
  order, 29–30, 81
  paper slip (*see* Card catalog)
  printed, 79
  search (*see* Search engine)
  vs. Shelving (*see* Shelving, vs. Catalog)
  systematic, 29–30
  by William Croswell, 4–5, 74–75, 77
Cataloging, 4, 39, 70–71, 79–80
  central, 96–97, 113
  efficient, 73, 79, 113
  history of, 47
  medieval, 155
  on paper slips, 43–44, 107
  rules, 44, 46
  secure, 169
  war, 174–175
Channel, 5, 24, 37
Citation index, 62
Classification, 10, 32, 78, 82, 165
  decimal (*see* System, decimal)
Cogswell, Joseph Green, 69, 78–79
Colbert, Jean-Baptiste, 175
Cole, George Watson, 152, 156
Commonplace. *See* Loci communes
Completeness, 9, 56, 118–119, 121–122, 133
Computer, 3, 7–8, 62, 106, 139–141
Connectivity, 63–64, 67
Conscript, 26, 28–29, 31, 56, 110, 124
Conscription number. *See* House number
Copy, 132
Copy error, 42–43, 177
Cross-reference, 81, 176. *See also* Excerpt, cross-references of
Croswell, William, 4, 69–78, 81–82, 140, 162–164
Cue. *See* Keyword
Cultural studies, basics of, 116
Curtius, Ernst Robert, 160
Cutter, Charles Ami, 82
Cutting up, 13–14, 44, 46, 73, 114
  paper, 3–4, 46
  paper slips, 73, 148

Daly, Lloyd W., 146
Database, 67, 111, 117
Data file, 106

Data processing, 1, 120–121, 139, 169
Davidson, Herbert E., 94–98, 100–102, 104, 112, 141, 167–169
Decimal system. *See* System, decimal
Denis, Michael, 40, 44
Deposit copy, 36
Desk, 3, 123, 129, 141
Dewey, Melvil, 5, 87–98, 102–104, 107, 121, 140, 150, 165–169
Dewey decimal classification system. *See* System, decimal
Disruption, 3, 5, 71–73, 111
Doni, Anton Francesco, 9
Drawer, 53–54, 80–81, 90–91, 98, 112, 170
Duration, 20, 42, 74, 131
 of the catalog manufacture, 82–83
 of the catalog use, 39, 44

Ebert, Friedrich Adolf, 28, 52, 152–153
Edwards, Edward, 88
Efficiency, 3, 73, 123, 175
Encyclopedia, 32, 64
Energetics, 116,123
Energy, 88, 117, 123
 saving, 117, 124–125
 waste of, 175
Enlightenment, 32–34, 62
Erman, Wilhelm, 156, 171, 174
Excerpt
 book, 17, 149
 cross-references of, 63–66
 organization, 50–55
 as pointer, 63
 processing, 13
 rereading, 55–56
Excerption, 4, 17, 149

Field, Herbert Haviland, 166
File, 178
Financial crisis, 60
Florilegia. *See* Excerpt
Flusser, Vilém, 179

Folsom, Charles, 79, 165, 169
Ford, Henry, 125
Fordism, 125
Form, 110
Format
 of a book, 10, 31, 44, 156
 of catalog slips, 82, 92, 166, 169
 index card (*see* Index card)
 international, 82, 118
 paper, 117
 playing card, 33
 standardized, 47, 79, 80, 117
 world (*see* World, format)
Foucault, Michel, 6, 146, 151
Fragmentation, 61
French Revolution, 45, 47
Froschauer, Christoph, 147

Genius, cult of, 62
Gessner, Konrad, 3, 9–10, 12–13, 16–17, 24, 39, 42, 47, 49, 68, 75, 114, 134, 140, 147–148, 163
Gilbreth, Frank B., 108, 124
Glance, 66
Gödel, Kurt, 121, 174
Goethe, Johann Wolfgang von, 57–58, 78, 160, 164
Goldsmith's note, 58
Gunn, James Newton, 169

Happiness, formula of, 120
Hardenberg, Friedrich von, 177
Hardware, 20, 82
Harris, Thaddeus William, 79–81, 164, 169
Harsdörffer, Georg Philipp, 21, 52
Hauptmann, Gerhart, 158
Hegel, Georg Wilhelm Friedrich, 49, 57, 159–160
Heidegger, Martin, 170
Heine, Heinrich, 57–58, 62–63
Hempel-Kürsinger, Johann-Nepomuk Freiherr von, 151

Herbarium, 50
Heyne, Christian Gottlob, 78
Hinz Büromaschinen GmbH, 133, 174
Historiography, 138
History, 129
  end of, 139
Hollerith, Herman, 101, 106, 136
Homer, 120
Hottinger, Christlieb Gotthold, 118–119
House number, 4, 27–28, 31, 58, 151
Hypertext, 65

IBM, 101, 106
Index, 12–13, 100. See also Register
  catalog, 149–150
  of forbidden books, 38
Index card, 6, 88, 92, 104, 106
  associative linkage through, 63, 66
  as an author's assistant, 62, 63
  circulation of, 62
  dimensions of the, 138
  format, 92, 100, 106, 110
  origin of, 47
  standardization of, 91–92
Indexing, 4, 13, 44, 47, 54, 74, 98, 102, 125, 130, 134
  as copying, 63
  genealogy of, 4
  loose, 17
  primal scene of, 1, 3, 6, 16, 69, 139
  severe, 69
  term, 145
Innovation, 64–65
Inspiration, 65–66
Instigation, etymology, 179
Institut International de Bibliographie, 92, 113, 126
Interface, 41, 46, 52, 66, 110
Interlocutor, 65–66
Inventory, 10, 17, 35

Jacquard, Joseph Marie, 106
Jansenism, 36
Jean Paul, 4, 53–57, 157, 159
Jesuits, 36–38, 43–44, 154
Jewett, Charles Coffin, 69, 80, 102–104, 169
Joseph II., 37
Jungius, Joachim, 17, 19, 20–21, 140, 148–149, 157

Kafka, Franz, 27
Kant, Immanuel, 66, 120
Kartothek (*see* Card index)
Kayser, Albrecht Christoph, 30, 152, 155, 158
Keyword, 10, 66, 123, 159–160
Kittler, Friedrich, 57, 146–147, 149, 177
Klinckowström, Carl von, 169
Kollektaneen. *See* Excerpt

Labor
  division of, 39, 47, 129
  mental, 117
Ladewig, Paul, 107, 109–110, 113, 140, 156
La Fontaine, Henry, 113–114, 119, 172
LeBlond, Gaspard-Michel, 46, 157
Lefèvre d'Ormesson, Louis-Francois-de-Paul, 45
Leibniz, Gottfried Wilhelm, 16, 19, 21–24, 30, 32, 39, 44, 51, 149–150, 152, 160, 165
Lessing, Gotthold Ephraim, 32–33, 51, 149, 152
Leyh, Georg, 148, 151–153
Librarian, 16, 89, 92, 98, 111, 138, 155
  activity of the, 30, 31, 33, 44
  ancient, 6
  assembly, 87, 103, 169
  change of office, 29
  Leibniz as, 21, 24, 29
  vs. scholar, 16, 50, 51
  studies, 4, 70

Library, 37, 40, 135
  access to (*see* Access)
  in Alexandria, 6
  Amherst College Library, 100
  architecture, 108
  arrangement, 28, 29
  auction, 149
  as battlefield, 153
  in Berlin, 108–109, 112, 123, 156
  vs. *Bücherei*, 135
  diary, 71, 75, 77
  and factory, 109
  female employees in the, 82
  fire at, 153, 165
  in Göttingen, 69, 79
  at Harvard, 68–72, 74–75, 77–78
  history, 10, 38, 70, 83, 88
  vs. *Kartothek*, 134–135
  Library of Congress, 88
  management, 30, 42, 90, 98, 153–154
  medieval, 9, 35
  monastery, 36–37
  new buildings of the, 108
  in New York, 80, 96
  vs. office, 5, 94, 103
  order, 30–32, 44, 142
  public, 171
  and scholarship, 16, 51
  science, 31, 69, 78
  servant, 155
  technology, 4, 5, 16, 67, 106, 108, 112–113
  universal, 32, 165
  in Vienna, 16
    city library, 37
    court library, 16, 21, 34–44, 111, 153
    university library, 43, 87
  in Wolfenbüttel, 16, 21, 32, 51
Library Bureau, 5, 83, 106, 108–110, 112–113, 131
  advertisement, 141
  bankruptcy, 98–99, 102, 104, 112, 118
  competition, 104
  expansion, 97, 98, 100, 101, 112
  foundation, 94
  gaining market, 95, 100
  manufacture of products, 100
Lichtenberg, Georg Christoph, 56
Liquidity, 61
List, 4, 10, 12, 31, 40, 42. *See also* Index
Location, 28
Loci communes, 9, 12–13, 62, 149
Lohenstein, Daniel Caspar, 149
Loose-leaf binder, 126
Luhmann, Niklas, 65, 143, 145, 157, 159, 173

Machine, 1, 129
  calculating, 129
  definition of the, 7
  and human, 129
  index (*see* Index)
  memory, 120
  office, 129, 131
  paper, 1, 3, 8, 99, 137, 140
  paper slip, 52
  scholar, 4, 50–52, 56, 63, 67, 157
  universal discrete (*see* Computer)
Mallarmé, Stéphane, 177
Marginal note, 173
Market value, of an author, 62
Maria Theresa, 27, 35–36, 154
Material, 6, 12–13
  adjustment of, 16, 47, 49, 54, 56–57, 73
  collection, 12, 52, 119, 121
  reviewing, 54
Mathetic. *See* Order, science of
Media change, 136
Memory, 31
  external, 66
  local, 28, 31, 52, 152
  machine (*see* Machine, memory)
  mobile, 128
  world (*see* World, memory)

Meyer, Richard M., 120, 162
Milkau, Fritz, 42, 165, 167
Mobility, 3, 8, 30–31, 42, 77, 128, 141, 170
 of paper slips, 13, 18, 163
 of records, 1, 117
 of tabs, 169
Modernity, 129
Monism, 120
Monographic Principle, 117
Moser, Johann Jacob, 4, 49, 53–54, 57, 61–62, 91, 158–159, 169
Müller, Heiner, 131–132
Murr, Christoph Gottlieb von, 150
Musil, Robert, 27

Nabokov, Vladimir, 67, 162
Naudé, Gabriel, 32
Neglect, 52, 157
Network, 61, 63–64
Nietzsche, Friedrich, 19, 144
Noise, 5
Normalization. *See* Standardization
Numbering, 28, 31

Office, 3, 5, 83, 93, 137
 and business trade show 1913, 122
 and card index, 90
 vs. library, 100
 machine (*see* Machine, office)
 organization, 5, 79–80, 124, 126, 131–133
  apparatus of the, 120
  journals for the, 126, 137
  reform, 108, 142
Order, 1, 9, 13, 30, 49–50, 55–56, 111, 133
 alphabetical, 9–10, 35, 44, 96, 104
 endangered, 42, 111
 catalog (*see* Catalog, order)
 rigid, 78
 science of, 125, 136

specific, 44
temporary, 74
Organization, 117, 125, 129. *See also* Office, organization
 efficient, 136
 of mental labor, 116
 as science, 123, 125, 137
 of studies, 24
 term, 176
Ostwald, Wilhelm, 116–117, 119–125, 136, 172–175, 179
Otlet, Paul, 113, 119

Panizzi, Anthony, 69
Paper, 33, 35
 vs. cardboard, 47, 80
 cutting up, 3–4, 46
 format, 117
 lack of, 158
 machine (*see* Machine, paper)
 money, 7, 59–62, 90, 160
  circulation, 59–62
  as currency, 59–60
  implementation, 58
  origin, 58, 60
  reputation, 58, 60–61
  as substitute, 60–61
 sheet, 53
 tape, 137, 139–140
Paper slips, 3, 21, 41, 45, 52–53, 81, 106, 110, 134
 bank, 7, 96–97
 cutting up (*see* Cutting up, paper slips)
 demolition of, 42–43, 157
 in Ladago, 158n12
 loose, 83
 loss of, 41, 44, 54, 95, 111, 155, 169
 modularity of, 52, 128–129, 170
 movement of, 6, 13, 44, 70
 print, 97, 112, 167
 rearrangement of, 44, 55, 74, 79, 108–109

Paper slips (*cont.*)
  safekeeping of, 14–15, 53–54, 74
  saving of, 75, 82, 104, 112, 171
  standardized, 80
Paper slip economy, 78, 80, 133, 140–141
Patenting, 104
Placcius, Vincentius, 17–18, 20, 52, 157
Playing card. *See* Card, playing
Pointer, 52, 118, 121
Polyhistorism, 75
Porstmann, Walter, 134–136, 138, 176–179
Portheim, Max von, 144
Production
  knowledge, 65
  textual, 62–63, 67
Productivity, 8, 35, 54, 62, 70–71, 75, 104, 113
Progress, 1, 79, 92, 99, 129, 141
  of cataloging, 41–42, 83
  Fortschritt Fabriken GmbH, 1, 137
Projector, 161
Psychotechniques, 125
Punch Card. *See* Card, punch

Quote, 10, 62

Rationalization, 124–125, 140, 168, 175
Rautenstrauch, Franz Stephan, 43, 44, 152, 156
Rearrangement, 53
Receiver, 3, 5, 106
Reference, 121
Register, 12–13, 49, 55–56, 70, 81, 149, 157
  ancient, 6
  manufacture, 17, 21
  universal, 114, 131
  vertical files, 100, 101
Registry. *See* Index
Remembrance, 102
Remington Rand, 106

Repository. *See* Bookshelf
Representation, 29
Reusability, 14–15, 81
Roesler, Rudolf, 123
Rozier, Abbé Francois, 33, 45–46, 80, 152

Saager, Adolf, 116, 119, 165, 172–173
Scheme, 11–12, 17, 28, 50, 115. *See also* Algorithm
Schlegel, Friedrich, 177
Schlesinger, Georg, 125, 175
Schmidt, Arno, 67, 146, 158, 162
Schmitt, Carl, 214
Scholar, 10, 12, 16–17, 49, 51–52, 57, 67, 138
Scholarship, 16, 63, 147, 157, 159
Schrettinger, Martin, 31, 152, 153
Schwenter, Daniel, 82, 164, 169
Search engine, 31–32, 34, 42, 47, 50
Sender, 3, 5
Shannon, Claude Elwood, 5
Shelving, 31, 79, 87, 149–150, 156
  vs. catalog, 5, 29–33, 42, 78
  and Enlightenment, 32–34
  mechanical, 30
  numerus currens, 30
  systematic, 30–32, 34, 42, 44
Sherman, E. W., 95, 168
Shortage of space, 6, 29, 108–109
Sibley, John Langdon, 71, 80, 82
Signature, 29–31, 40–41, 123, 151, 152
Slip box. *See* Card index
Slips. *See* Paper slips
Sloth, 5, 68, 73, 140
Smithsonian Institute, 79, 104
Soennecken, Friedrich, 126, 171
Software, 8, 20, 82
Soldier. *See* Conscript
Solvay, Ernest, 117
Spelling reform, 87–90, 165
Staderini, Aristide, 171

Standardization, 15, 47, 79, 81, 90, 91, 110, 124–126
Storage, 1, 3, 20, 52, 90, 99, 120
  recursive, 173
Stream, 5
  of books, 31, 43
  data, 35, 37, 41, 47, 56
Surprise, 64
Swieten, Gerhard van, 35–36, 38, 153–154
Swieten, Gottfried van, 34, 36, 38–40, 42, 153–154
Swift, Jonathan, 158
System, 17
  administrator, 128
  anthropometric, 101
  of book shelves, 109
  complete, 120
  decimal, 88, 113, 165
  economic, 96
  of economic exchange, 61
  flexible, 13, 77
  magazine, 126–127
  metric, 87–88, 90, 117, 165
  multiuser, 4
  promissory note, 61
  of saving paper slips, 111–112
  standardized, 39
Systematics, 30, 32–33, 74, 88, 150, 152, 156, 165
  genealogy of, 152
  of war, 174–175

Tab, 94, 169
Table, 33, 46, 152
Taylor, Frederick Winslow, 108, 123–125, 175
Taylorism, 15, 108, 123–125
Technology transfer, 69, 94, 99, 124–125
Time, 110
  saving, 89, 90, 93, 97, 99, 102, 118, 125, 127, 165

Totality, 9, 57, 118–119, 121–122, 133
Transfer, 3–6, 89, 104, 136, 140
  of books, 40
  information, 5, 6, 41, 44
  of library technique, 4, 67–68, 104, 131
Turing, Alan Mathison, 1, 8, 148–149
Type case, 14–16
Typewriter, 3, 101, 106, 126

Vaucanson, Jacques de, 106
View, 66
Vogt, Victor, 137–138, 179
Voucher, 58, 59

Weaving, Moser's technique of, 61, 161
Witte, Irene Margarete, 124, 175
World, 114, 117–118
  archive for humor, 120
  brain, 114
  catalog, 113
  encyclopedia, 117
  exhibition, 101
  format, 113, 117, 121–123, 173
  knowledge, 117
  memory, 113
  postal union, 92
  registry, 121
  war, 122–124, 127
War
  economy, 110, 124, 127
  game, 56
  world, 122–124, 127
Work studies, 5, 108, 125, 128
Writing / Reading head, 1, 3, 8, 40, 52, 140
Writing, three-dimensional, 138

www.ingramcontent.com/pod-product-compliance
Lightning Source LLC
Chambersburg PA
CBHW020836020526
44114CB00040B/1220